Platt Sawyer

Catalogue of the library of the late Platt R.H. Sawyer, M.D. of Bedford, Westchester County, N.Y.

Platt Sawyer

Catalogue of the library of the late Platt R.H. Sawyer, M.D. of Bedford, Westchester County, N.Y.

ISBN/EAN: 9783741194047

Manufactured in Europe, USA, Canada, Australia, Japa

Cover: Foto ©Andreas Hilbeck / pixelio.de

Manufactured and distributed by brebook publishing software (www.brebook.com)

Platt Sawyer

Catalogue of the library of the late Platt R.H. Sawyer, M.D. of Bedford, Westchester County, N.Y.

CATALOGUE.

The sizes of books in this catalogue are defined relatively to signatures and according to the following page, height and inch measurements :
LARGE FOLIO, over 18 inches ; FOLIO, below 18 and over 13 ; SMALL FOLIO, below 13 and over 11.
LARGE QUARTO, below 15 and over 11 ; QUARTO, below 11 and over 8 ; SMALL QUARTO, below 8 and over 6.
LARGE OCTAVO, below 11 and over 9 ; OCTAVO, below 9 and over 8 ; SMALL OCTAVO, below 8 and over 6.
TWELVEMO, below 8 and over 6. MINIMO, below 6 inches.

1 ACTRESS in High Life ; COLTON's Ship and Ashore ; HEMANS' Poems; HOBART's Apostolic Order; etc. Together 10 vols. Cloth.

2 ADAMS (R. C.). Travels in Faith. Small 8vo, new cloth.
N. Y., 1884

3 ADAMS' Cross in Cell; Iceland, Greenland and Faroe Islands; THACKERAY's Men's Wives; DAUDET's New Don Quixote; SPARKS's BENEDICT ARNOLD; GEORGE SAND's Teverino; TICE's Kansas and Colorado; etc. Together 15 vols. Cloth.

4 AINSWORTH's Latin Dictionary; LEVERETT's Latin Lexicon; MORSE's Universal Gazetteer; HEDERICUS's Greek Lexicon; WEST's Bible Analysis. Together 5 vols. Full leather.

5 ALCOCK (Rutherford). Three Years in Japan. *Maps and illustrations.* 2 vols. 12mo, cloth. N. Y., 1868

6 ALDEN (J.). The Religious Life. Small 4to, cloth.
N. Y., 1879

7 ALEXANDER (Capt. J. E.). Trans-Atlantic Sketches in North and South America and the West Indies. *Plates.* 2 vols. 8vo, half calf, gilt (some pp. foxed). London, 1833

8 ALICE (Grand Duchess of Hesse). Biographical Sketch and Letters. *With portrait.* Small 8vo, cloth. N. Y., 1885

9 ALICE. Another copy of the same.

10 ALISON (Sir Archibald, *LL.D.*). History of Europe from 1789 to 1852, including Continuation and Index Volume. *Illustrated.* Together 21 vols. Small 8vo, cloth, uncut.
Edinburgh, 1855–70

11 ALISON. Essays Political, Historical and Miscellaneous. 3 vols. thick 8vo, cloth, uncut. Edinburgh, 1850

12 ALTAR SERVICE of P. E. Church. Large 8vo, roan, gilt (rubbed). Phila., *n. d.*

13 AMERICAN CHURCH REVIEW, Nos. 132 to 135 inclusive, and Nos. 137, 138 and 139, 7 vols. cloth; The Same, Nos. 143 to 149 inclusive, also No. 151, 8 parts, sewed. Together 15 pieces. 8vo. N. Y., 1881–83

14 AMERICAN HOME BOOK. Thick small 8vo, new cloth. N. Y., *n. d.*

15 AMICIS (Edmondo de). Works, *i. e.*:—Military Life in Italy, *illustrated;* Constantinople; Morocco, its People and Places, *illustrated;* Studies of Paris; Spain and the Spaniards, *illustrated.* Together 5 vols. Small 8vo, fresh uniform cloth. N. Y., 1882–85, etc.

16 AMOURS DE ZEOKINZUL ROY DES KOFIRAINS. Ouvrage traduit de l'Arabe du Voyageur KRINELBOL. 12mo, boards. Amsterdam, 1764
Erotic and rare.

17 ANDERSON (Alex. D.). Silver Country, or the Great Southwest. Small 8vo, new cloth. N. Y., 1877

18 ANDERSEN (Hans Christian). Stories for the Household. *With 200 illustrations.* Thick small 8vo, cloth.
London, *n. d.*

19 ARVINE (Kazlitt). Cyclopedia of Moral and Religious Anecdotes. *Portrait.* Thick large 8vo, morocco, gilt edges. N. Y., 1881

20 ATKINSON (Edward). Distribution of Products. Small 8vo, fresh cloth. N. Y., 1884

20* ATKINSON. Another copy of the same.

21 ATKINSON (T. W.). Travels in Upper and Lower Amoor. *Map and illustrations.* Large 8vo, cloth. N. Y., 1860

22 ATLANTIC MONTHLY, 95 parts, of which the years 1866, '67 and '68 are complete. Large 8vo, sewed.
Boston, 1862–77

23 AUDSLEY (W. *and* G.). Polychromatic Decoration as applied to Building in the Mediæval Styles. *Illuminated plates.* Folio, cloth, top edge gilt.

London, *Henry Sotheran & Co.*, 1882
This work comprises thirty-six folio plates, executed in the highest style of chromolithography, in colors and gold, with descriptive letterpress. The introductory text contains a brief historical essay and practical hints on the processes of painting—oil, tempera, and wax painting—preparing the design, and transferring the same to the walls or other portions of a building, and the principles of decorative design and coloring.

24 AUDSLEY. Another copy of the same. (Back slightly damaged.)

25 AUTHORS AND PUBLISHERS—Suggestions for Beginners in Literature. Large 8vo, new cloth. N. Y., 1884

26 AVILLION and other Tales, 3 vols.; WM. WILBERFORCE'S Life ; KEITH's Signs of Times ; etc. Together 10 vols. Cloth.

27 BACON (M. A.). Winged Thoughts. *With charming illuminated pages, colored plates of birds, printed in gold and tints—illuminations designed by* OWEN JONES *and drawn on stone by* BATEMAN. Large 8vo, calf, bevelled sides, gilt edges. London, 1851

28 BALLADS OF BEAUTY, edited by BARKER. *With* 40 *page illustrations.* 4to, boards. Boston, 1874

29 BANCROFT (George). History of the United States. *Portraits, etc,* 10 vols. large 8vo, cloth. Boston, 1874

30 BAPTIST PRAISE BOOK. 4to, morocco, gilt edges (rubbed). N. Y., 1871

31 BARDSLEY (Charles W.). Curiosities of Puritan Nomenclature. Small 8vo, cloth, top edge gilt. N. Y., 1880

32 BARHAM (R. H.). Ingoldsby Legends of Mirth and Marvels. *Illustrated by* CRUIKSHANK, LEECH, TENNIEL *and* BARHAM. Small 8vo, cloth, uncut. London, 1882

33 BARNARD (Charles). Co-operation as a Business. Square small 8vo, new cloth. N. Y., 1881

34 BARTH (H.), African Travels ; LIVINGSTONE's Travels in South Africa; JACKMAN, the Australian Captive, *illustrated.* Together 3 vols. Cloth.

35 BEATTIE (William). Castles and Abbeys of England. *With upwards of* 200 *steel and wood engravings.* Large 8vo, cloth, gilt edges.   London, n. d.

36 BEAUMONT (A.). History of Spain. *Front.* 8vo, calf (foxed).   London, 1809

37 BENSON (S.). My Visit to the Sun. 8vo, cloth. N. Y., 1874

38 BENTON (Thos. H.). Thirty Years' View. *Portrait and view.* 2 vols. large 8vo, cloth.   N. Y., 1875

39 BESTE (J. R. D.). Nowadays, or Courts, Courtiers, Churchmen, Garibaldians, Lawyers and Brigands. 2 vols. 8vo, cloth, uncut.   London, 1870

40 BIRD (Isabella L.). The Golden Chersonese and the Way Thither. *With map and illustrations.* Small 8vo, new cloth.   N. Y., 1884

41 BISHOP (J. Leander). History of American Manufactures from 1608 to 1860. *Portraits.* Vols. 1 and 2. Thick 8vo, cloth.   Phila., 1866

42 BISHOP (N. H.). Four Months in a Sneak-Box, a Boat Voyage of 2,600 Miles. *Maps and illustrations.* 8vo, cloth.   Boston, 1879

43 BLACKWOOD'S MAGAZINE. 95 parts, of which years 1863, '66, '67 and '68 are complete. Together 95 parts. Large 8vo, sewed.   N. Y., 1863–70

44 BLOOMFIELD (Lady Georgiana). Reminiscences of Court and Diplomatic Life, 1841–1870. By LADY BLOOMFIELD. *With portraits on steel and other illustrations.* 2 vols. 8vo, half calf.   London, 1883
"These amusing and interesting memoirs."—*N. Y. Herald.*

44* BLOOMFIELD. The Same. 2 vols. 8vo, fresh cloth.   N. Y. [London], 1883

45 BOCCACCIO. Decameron, with Introduction by WRIGHT. *With the Milan and* STOTHARD *plates.* Thick small 8vo, cloth.   London, n. d.

45* BOCCACCIO. Decameron. Small 8vo, sewed. N. Y., n. d.

46 BORROW (George). The Bible in Spain. 3 vols. small 8vo, cloth, uncut.   London, 1843

47 BORROW. Lavengro; the Scholar, the Gypsy, the Priest. *Portrait (foxed).* 3 vols. small 8vo, cloth, uncut.   London, 1851

48 BORROW. The Romany Rye, a sequel to "Lavengro." 2 vols. small 8vo, cloth, uncut (binding damaged).   London, 1858

49 BOWLES (J. L.). Japanese Marks and Seals. Part I.,
Pottery. Part II., Illuminated MSS. and Printed Books.
Part III., Lacquer, Enamels, Metals, Wood, Ivory, etc.
*Comprising 1,300 marks and seals copied in fac-simile,
with examples in colors and gold, executed by* MESSRS.
FIRMIN-DIDOT, *of Paris.* Thick large 8vo, cloth, top
edge gilt.

London *and* Manchester, *Henry Sotheran & Co.*, 1882
The work also contains a grammar of the Marks ; brief histor-
ical notices of the various arts of Japan ; the Jikkwan and Jinni
Shi Characters, with those of the Zodiacal Cycle ; also the Year
Periods since the Fourteenth Century ; and a Map showing the
various seats of manufacture.

50 BOWRING (John). Specimens of the Russian Poets. 12mo,
half calf, uncut (name on title). London, 1821

51 BOYDELL (John). Gallery of Illustrations for SHAKESPEARE'S
Dramatic Works, with Selections from the Text, edited
by J. PARKER NORRIS. *Illustrated with plates, reduced
and re-engraved by the heliotype process, and selected
from the originals published by* JOHN BOYDELL. Large
4to, cloth, gilt edges. Phila., n. d.

52 BRASSEY (Sir Thomas). Work and Wages. Square small
8vo, cloth. N. Y., 1883

53 BRITISH QUARTERLY; Nineteenth Century ; etc. Together
19 parts. Sewed.

54 BROWN (Edmund Woodward). Life of Society. Thick
8vo, fresh cloth. N. Y., 1885

55 BROWNELL (T. C., *Bishop of Connecticut*). Family Prayer
Book. Thick 4to, calf. N. Y., 1850

56 BRUCE (Charles). Book of Noble Englishmen. *Front.,*
(*group of portraits*). Small 8vo, half red morocco gilt,
cloth sides, marbled edges. Edinburgh, 1880

57 BRYANT (Wm. Cullen). Three Great Poems—Thanatopsis,
Flood of Years and Among the Trees. *Illustrated by*
LINTON *and* McENTEE. 4to, cloth gilt, edges gilt.
N. Y., 1877

58 BRYANT HOMESTEAD BOOK. By "The Idle Scholar"
[JULIA HATFIELD]. *With portraits of* BRYANT *by*
SARONY *and* NAST, *and 18 engravings on wood, drawn
by* Hows, *engraved by* LINTON. 4to, cloth, gilt edges
(title upside down). N. Y., 1870

59 BRYANT HOMESTEAD BOOK. Another copy.

60 BRYANT. Among the Trees. *Illustrated by* McENTEE.
4to, fresh cloth.   N. Y., 1884

61 BRYANT. Thanatopsis. *Illustrated by* LINTON. 4to, new
cloth, gilt edges.   N. Y., n. d.

62 BUCKLE (H. T.). History of Civilization in England. 2
vols. small 8vo, cloth.   N. Y., 1876

63 BUFFON (Le Comte de). Œuvres d'Histoire Naturelle.
Numerous plates. 40 vols. small 8vo, boards.
    Berne, 1792

64 BUFFON. Œuvres. Vols. 1 to 6 inclusive, also 9 to 13 inclusive. *Portrait and plates.* Together 11 vols. 12mo,
boards.   Paris, 1774–78

65 BUILDER (The) for 1869. *Illustrations.* Thick small folio,
half roan (no title).   London, 1869

66 BUNYAN (John). Pilgrim's Progress, edited by PHILIP.
With numerous steel and wood engravings. Thick
large 8vo, cloth.   N. Y., 1882

67 BUNYAN. The Same. *Illustrated.* Thick 8vo, cloth.
    N. Y., 1880

68 BUNYAN. Works in Welsh—*i. e.*, Taith y Pererin ; y Rhy-
fel Ysprydol; a Bywyd a Marwolaeth Mr. Drygddyn; etc.
Colored plates. Thick large 4to, roan gilt, bevelled sides,
gilt edges.   Caerdydd, n. d.

69 BURNING WORDS of Brilliant Writers—a Cyclopædia of
Quotations by GILBERT and ROBINSON. Thick 8vo, cloth.
    N. Y., 1883

70 BURNS (Robert). Complete Poetical and Prose Works,
with Life, Notes, Correspondence and Glossary by A.
CUNNINGHAM. *Illustrated.* Thick large 8vo, cloth (back
inked).   N. Y., 1881

71 BURTON (John Hill). The Book-Hunter. *Portrait and
view.* Small 8vo, half russia, totally uncut (binding damaged).   N. Y., 1883

72 BUSH (C. G.). Our Choir, a Symphony in A, B, C, D, E,
F, G, etc., Flat and Sharp, Major or Minor. Oblong
folio, boards.   N. Y., 1883

    "It is a very laughable thing, and withal containing some whole-
some hints that the choir-singers will best appreciate. No choir
should be without one. It will most pleasantly relieve the tedium
of a long, dry sermon."—*Rochester Herald.*

73 BUTTERWORTH (J.). Scripture Concordance. 8vo, cloth.
Phila., n. d.
74 BYRON (Lord). Poetical Works. *Illustrated.* Thick
large 8vo, cloth. N. Y., n. d.
75 BYRON. Childe Harold's Pilgrimage and other Poems.
Small 8vo, half morocco, carmine edges. London, 1814

76 CÆSAR. Commentaries. Literal Translation. 12mo,
cloth. N. Y., 1875
77 CAMERON (V. L.). Across Africa. *With numerous illustrations.* 2 vols. 8vo, half calf gilt, marbled edges.
London, 1877
78 CAMPBELL (Helen). American Girl's Home Book of Work
and Play. *Illustrated.* Small 8vo, cloth, gilt.
N. Y., 1883
79 CAMPION (J. S., *Major U. S. A.*). On the Frontier—Wild
Sports, Personal Adventures and Strange Scenes. *Illustrated.* 8vo, cloth (binding soiled). London, 1878
80 CAMPION. Another copy of the same (binding damaged).
81 CARTER (S. N.). Art Suggestions from the Masters. Small
8vo, fresh cloth, red top edge. N. Y., 1881
Includes—Sir Joshua Reynolds, Sir Charles Bell, William Haz-
litt and Benjamin R. Haydon.
82 CARLYLE (Thos.). WORKS. *Portrait.* 30 vols. small
8vo, cloth, uncut. London and N. Y., 1871, *etc.*
Including—French Revolution, 3 vols.; Cromwell's Life and
Letters, 5 vols.; Miscellaneous Essays, 7 vols.; Frederick the
Great, 10 vols.; Wilhelm Meister, 3 vols.; Sartor Resartus; Heroes.
83 CELEBRATED TRIALS AND REMARKABLE CASES
of Criminal Jurisprudence, from the Earliest Records to
the year 1825. *Plates.* 6 vols. small 8vo, half calf gilt.
London, 1825
84 CERVANTES (M. De). Adventures of Don Quixote, translated by JARVIS. *Illustrated by* TONY JOHANNOT. Thick
8vo, cloth. N. Y., n. d.
85 CHAMBLIN (J. E.). Lives and Travels of LIVINGSTONE and
STANLEY. *Profusely illustrated.* Thick 8vo, cloth.
Phila., 1881
86 CHAMBERS (W. *and* R.). Information for the People.
Illustrated. 2 vols. thick large 8vo, cloth. London, 1880

87 CHAMBERS's Etymological English Dictionary. Small 8vo,
cloth. London, 1883

88 CHASE (S. P., *U. S. Chief Justice*). Life and Public Services of, by J. W. SCHUCKERS. *Portrait.* Thick 8vo, cloth. N. Y., 1874

89 CHAVASSE (P. H.). Physical Training of Children. *Front.* 8vo, cloth. Phila., 1879

90 CHRISTMAS IN ART AND SONG—Songs, Carols and Descriptive Poems. *Illustrated from drawings by distinguished artists.* Cloth, gilt edges. N. Y., 1880

91 CHRISTMAS. The Same. Another copy.

92 CHRISTY's Cotton is King; GILBERT's Monomaniac; MULOCK's Woman's Thoughts about Woman; REDPATH's Hayti; Cross and Shamrock; etc. Together 15 vols. Cloth.

93 CIVIL ENGINEER and Architect's Journal. Vol. 29 [1866]. *Illustrated.* 4to, cloth. London, 1866

94 CIVIL WAR MAPS—Battle of Gettysburg, 3 pieces; Atlanta Campaign, 6 pieces; Knoxville, Tenn., 1 piece; Chattanooga, 1 piece; Franklin, Tenn., 1 piece; Sherman's Campaign, 1 piece. (13 pieces.)

95 CLARK (Willis Gaylord). Literary Remains. Large 8vo, cloth (foxed). N. Y., 1855

96 [CLEMENS.] Prince and Pauper, by "Mark Twain." *With 192 illustrations.* 4to, cloth (back rubbed). Boston, 1882

97 CLEVELAND (Grover). Life and Public Services, by PENDLETON KING. *Steel portrait and cuts.* Small 4to, cloth. N. Y., 1884

98 CLINTON (H. F.). Fasti Hellenici—the Civil and Literary Chronology of Greece and Rome. Thick large 4to, cloth, uncut. Oxford, *University Press*, 1851

99 COCHRANE (A. B.). Theatre Français in Reign of Louis XV. 8vo, fresh cloth. London, 1875

100 COLANGE (L. de, *LL.D.*). National Gazetteer of the U. S. Thick large 8vo, sheep, marbled edges. N. Y., 1884

101 COLERIDGE (S. T.). Letters, Conversations, and Recollections of. 2 vols. 12mo, half calf, gilt. ORIGINAL EDITION. London, *Moxon*, 1836

102 COLERIDGE. Poetical Works. *Illustrated and red borders.* Small 8vo, cloth, gilt edges. N. Y., n. d.

103 COLLINS (William). Poetical Works, with Life by JOHNSON, Notes by DYCE, etc. Small 8vo, cloth, uncut. LARGE PAPER. London, *W. Pickering*, 1827

104 COLMAN (George, *the younger*). Broad Grins. *Illustrated.* Small 8vo, cloth, uncut (cover loose). London, *n. d.*

105 CONGRESSIONAL MEMORIAL ADDRESSES on ANDREW JOHNSON, HENRY WILSON, MICHAEL C. KERR, J. E. LEONARD, E. Y. PARSONS, FRANK WELCH, GUSTAVE SCHLEICHER, A. S. WILLIAMS, T. J. QUINN, JULIAN HARTRIDGE, B. R. DOUGLAS, W. P. FESSENDEN, and O. S. FERRY. *Portraits.* Together 13 vols. Large 8vo, cloth. Washington, 1870–80

106 COPPINGER (R. W.). Cruise of the "Alert," Four Years in Patagonian, Polynesian and Mascarene Waters, 1878–82. *With* 16 *full-page woodcuts.* 4to, cloth, gilt edges. N. Y., 1884

107 COULTHARD (H. C.). Blast Engines, with letter-press description. *With plates to scale of large working drawings.* Large folio, half roan, cloth sides. London, 1867

108 COX (David). Biography, with Remarks on his Works and Genius by WM. HALL. *Portrait.* 8vo, cloth. N. Y., n. d.

109 COX (Samuel S.). Orient Sunbeams and Arctic Sunbeams. *Illustrated.* 2 vols. small 8vo, fresh cloth. N. Y., 1882

110 COX. Search for Winter Sunbeams. *With numerous illustrations.* Small 8vo, cloth. N. Y., 1870

111 CRAIK (G. L.). Manual of English Literature. Small 8vo, cloth, uncut. London, n. d.

112 CRÉBILLON. Œuvres avec Vie. *Portrait by* LEMOINE. 3 vols. minimo, calf, gilt edges (rubbed). EDITION CAZIN. Paris [Londres], 1785

113 CRUIKSHANK. 1851, or the Adventures of Mr. and Mrs. Sandboys at the Great Exhibition by HENRY MAYHEW. *With etched plates by* GEORGE CRUIKSHANK. 8vo, half calf, gilt. London [1851]

114 CRUIKSHANK.—BERENGER (Baron de). Helps and Hints how to Protect Life and Property. *With illustrations by* G. *and* R. CRUIKSHANK *and others.* 8vo, half morocco, gilt. London, 1835

115 CUMMING (R. G.). Hunter's Life in Africa. 2 vols. 12mo, cloth. N. Y., 1856

116 CUNNINGHAM (Peter). The Story of NELL GWYN and the Sayings of CHARLES THE SECOND. UNIQUE COPY, EXTRA ILLUSTRATED *with a number of steel engravings of portraits*, PROOFS BEFORE LETTERS. Large 8vo, half blue crushed levant morocco gilt, top edge gilt, others uncut. N. Y., 1883

117 CUNNINGHAM. The Same. UNIQUE COPY. Large 8vo, half red crushed levant morocco gilt, top edge gilt, others uncut.

118 CUNNINGHAM. The Same. UNIQUE COPY. Large 8vo, half calf gilt, top edge gilt, others uncut.

119 CURRAN (John Philpot). Life of, by his son W. H. CURRAN. *Portrait.* 12mo, cloth. Chicago, 1882

120 CYRILLA by BARONESS TAUTPHOEUS, 3 vols.; COX and HOBY'S Narrative; etc. Together 17 vols.

121 DAFFORNE (James). Albert Memorial, Hyde Park, its History and Description. *With numerous illustrations engraved on steel.* Large 4to, cloth, gilt edges. London, *n. d.*

122 DANA (Charles A.). Household Book of Poetry, Collected and Edited by DANA. Eleventh Edition Revised and Enlarged. *With illustrations.* Thick 4to, green cloth gilt, edges gilt. N. Y., 1878

123 DANA. Household Book of Poetry—eleventh edition. Revised and Enlarged. *Illustrated.* 4to, cloth, red edges. N. Y., 1882

124 DAVIDS (T. W. Rhys). The Origin and Growth of Religion, as Illustrated by Indian Buddhism, being the HIBBERT Lectures. Large 8vo, new cloth, uncut. N. Y., 1882

　　"Mr. Davids' lectures are an interesting and authoritative popular exposition of the points he discusses, written in a clear and graphic style."—*N. Y. World.*

125 DAVIDS. Another copy of the same.

126 D'AUBIGNÉ (J. H. M.). History of the Great Reformation. Vols. 1 to 5 inclusive. 8vo, cloth. London, 1843–53

127 DAY (Henry). From the Pyrenees to Pillars of Hercules. *Front.* 12mo, cloth. N. Y., 1883

128 DE FOE (Daniel). Works. *Portrait, and vignette title.* Large 8vo, cloth, top edge gilt. Brooklyn and N. Y., *n. d.*

129 DE QUINCEY (Thos.). Works, *i. e.*:—Historical and Critical Essays, 2 vols.; Narrative Papers, 2 vols.; Autobiographic Sketches ; Literary Reminiscences, 2 vols. Together 7 vols. 12mo, cloth. Boston, 1853-54

130 DE VERE (Schele). Wonders of the Deep. *Illustrations.* 12mo, fresh cloth. N. Y., 1885

DICKENSIANA, INCLUDING ORIGINAL EDITIONS.

131 DICKENS (Charles). Life and Adventures of Martin Chuzzlewit. *With full-page illustrations by* "Phiz." Thick 8vo, calf gilt (cover loose, foxed somewhat, and name on title). London, 1844
First edition and very scarce.

132 DICKENS. Life and Adventures of Nicholas Nickleby. *With illustrations by* "Phiz" *and portrait of author.* Thick 8vo, calf gilt (name on title). London, 1839
Original edition and very scarce.

133 DICKENS. Personal History of David Copperfield. *With plates by* H. K. Browne. Thick 8vo, cloth, uncut. London, 1850
Very scarce. Original large type illustrated edition.

134 DICKENS. Bleak House. *With illustrations by* H. K. Browne. Thick 8vo, cloth, uncut (cover loose). London, 1853
Very scarce. Original large type illustrated edition.

135 DICKENS. Little Dorrit. *With illustrations by* H. K. Browne. Thick 8vo, cloth, uncut (binding loose). London, 1857
Rare. Original large type illustrated edition.

136 DICKENS. Works, *i. e.*:—Pickwick Papers ; Nicholas Nickleby ; Martin Chuzzlewit ; David Copperfield ; Bleak House ; Mutual Friend ; Dombey and Son ; Little Dorrit ; Christmas Books, etc.; Curiosity Shop and Hard Times ; Child's England, etc.; Oliver Twist, Italy and America; Barnaby Rudge and Edwin Drood ; Great Expectations and Uncommercial Traveller; Tale of Two Cities and Boz Sketches. *Profusely illustrated.* Together 15 vols. Small 8vo, half red calf, marbled edges. N. Y., 1884, etc.

136* DICKENS. Works. The Same Edition. 15 vols. small 8vo, half purple calf, marbled edges.

137 DICKENS. Works, *i. e.*:—Bleak House; Two Cities and Boz Sketches; Great Expectations and Uncommercial Traveller; Barnaby Rudge and Edwin Drood; Nicholas Nickleby; Old Curiosity Shop and Hard Times; Mutual Friend; Pickwick Papers; Martin Chuzzlewit; Child's England and Miscellaneous. *Illustrated.* Together 10 vols. Small 8vo, half red russia (one back damaged). N. Y., n. d.

138 DICKENS. Letters of, 1833–1870. 3 vols. 8vo, cloth. London, 1880–82

139 DICKENS. Letters of, 1833–1870. 3 vols. small 8vo, cloth. N. Y., 1879–81

140 DICKENS. Pickwick Papers. *With* BARNARD's *illustrations.* Small 8vo, half calf, gilt. N. Y., n. d.

141 DICKENS. Nicholas Nickleby. *Illustrated.* Small 8vo, half calf. N. Y., n. d.

142 DICKENSIANA.—PICKWICK ABROAD, or the Tour in France, by GEORGE W. M. REYNOLDS. *Illustrated with steel engravings by* "CROWQUILL," *and woodcuts.* Thick 8vo, new cloth. London, 1864

143 DISRAELI (Isaac). Works, *i. e.*:—Curiosities of Literature, 3 vols.; Calamities and Quarrels of Authors; Amenities of Literature; Literary Character of Men. of Genius. *Portrait.* Together 6 vols. Small 8vo, fresh cloth. N. Y., 1880–81

144 DISRAELI. Curiosities of Literature, including American, by GRISWOLD. Large 8vo, sheep, marbled edges. N. Y., 1881

145 DISRAELI. The Same Edition. Cloth.

146 [DIXON (Henry Hall)] *i. e.*, "THE DRUID." Silk and Scarlet; Saddle and Sirloin; Post and Paddock; SCOTT and SEBRIGHT. *With steel engravings.* 4 vols. small 8vo, cloth gilt, others uncut. London, n. d.

147 DORÉ GIFT BOOK, with Introductory Notice of the Arthurian Legends. *With full-page illustrations to* TENNYSON's "*Idylls of the King,*" *by* GUSTAVE DORÉ. Large 4to, cloth gilt, edges gilt (cover loose). London, n. d.

148 DORÉ. Story of Arthur and Guinevere, and Fate of Sir Lancelot of the Lake, as told in Antique Legends and Ballads and in Modern Poetry. *With 9 illustrations by* GUSTAVE DORÉ. Large 4to, cloth, gilt edges (cover loose). London, n. d.

149 DORÉ. Story of Merlin and Vivien. *With 9 illustrations by* GUSTAVE DORÉ. Large 4to, cloth, gilt edges (some pp. loose). London, *n. d.*

150 DORÉ GALLERY of Bible Stories, with Descriptive Text by JOSEPHINE POLLARD. *Portrait and numerous illustrations by* DORÉ. Large 4to, half roan, cloth sides. N. Y., 1882

151 DOW (Lorenzo). Life, Travels, Labors and Writings. 2 *portraits.* Large 8vo, sheep. N. Y., 1881

152 DOWIE DENS O' YARROW. *With steel plates by eminent engravers after the original designs of* J. NOEL PATON, R.A. Folio, cloth.
Edinburgh, *for the Members of the Royal Association for Promotion of Fine Arts in Scotland,* 1860

153 DOWLING (John, *D.D.*). History of Romanism. *Profusely illustrated.* Thick large 8vo, cloth, marbled edges. N. Y., 1871

154 DOWNING (A. J.). Landscape Gardening. *Portrait and illustrations.* Thick large 8vo, cloth. N. Y., 1865

155 DRYDEN (John). Poetical Works. *Illustrated.* Small 8vo, cloth, gilt edges. London, *n. d.*

156 DUMAS (Alexander). Vicomte de Bragelonne, 2 vols.; Three Musketeers ; Twenty Years After. *Illustrations.* Together 4 vols. Small 8vo, cloth. London, 1878

157 DUMAS. Memoirs of a Physician. 2 vols. in 1. Small 8vo, half roan. London, n. d.

158 DUMAS (A., *fils*). Vie à Vingt Ans. Small 8vo, half morocco. Paris, 1865

159 EASTLAKE (Sir Charles). Pictures by, with a Biography and Critical Sketch of the Artist by W. COSMO MONKHOUSE. *Steel engravings by eminent engravers.* Large 4to, cloth gilt, edges gilt. London, n. d.

160 EDINBURGH REVIEW. Jan., 1863, to July, 1874, inclusive. Together 40 parts. Large 8vo, sewed. N. Y., 1863–74

161 EDWARDES (Herbert B.). Year on Punjab Frontier in 1848–49. *Portrait, colored plates, fac-similes, etc.* 2 vols. 8vo, half calf, gilt. London, 1851

162 ELIOT (George).   Works, *i. e.:*—Felix Holt; Silas Marner,
Lifted Veil, Brother Jacob, Scenes of Clerical Life;
Theophrastus Such, Legend of Jubal and other Poems,
Spanish Gypsy.   5 vols. in 3.   Small 8vo, half calf, gilt.
N. Y., n. d.
163 ELMENDORF (J. J.).   Lectures on History of Philosophy.
12mo, cloth.   N. Y., 1876
164 ELTON (J. F.).   Travels and Researches among the Lakes
and Mountains of Eastern and Central Africa.   *With
maps and illustrations.*   8vo, cloth, gilt edges.
London, n. d.
165 ENGLISH AS SHE IS SPOKE.   Minimo, new cloth.   N. Y., 1884
166 ENGLISH PHILOSOPHERS—BACON, SHAFTESBURY and HUTCH-
ESON, by THOMAS FOWLER; ADAM SMITH, by J. A. FAR-
RAR; DAVID HARTLEY and JAMES MILL, by G. S. BOWER.
Together 4 vols.   Square small 8vo, cloth.   N. Y., 1881-83

ENGRAVINGS, ETCHINGS, ETC.

167 PORTRAITS of Distinguished Americans—Presidents,
Statesmen, Generals, Authors, etc.   (50)
168 PORTRAITS.   A similar collection.   (34)
169 PORTRAITS.   Principally of American Generals, Statesmen, etc.   (166)
170 PORTRAITS of European Celebrities—Kings, Queens,
Statesmen, Generals, etc.   (69)
171 MISCELLANEOUS VIEWS—English, Continental, Eastern, etc.
(75)
172 SCRIPTURE ILLUSTRATIONS, after POUSSIN, WEST, etc.   (30)
173 ILLUSTRATIONS to the Book of JOSHUA, after MELVILLE (33
duplicates) and SINGLETON (21 duplicates).   (54)
174 PHOTOGRAPHS—Portraits, Views, etc.   (27)
175 SHAKESPEARE—Views, etc., suitable for illustrating
the Works of WILLIAM SHAKESPEARE.   (27)
176 ENGRAVINGS—Various, after WATTEAU, LEUTZE, etc.   (50)
177 ENGRAVINGS—Various, after DORÉ, LOUTHERBOURG, etc.,
some colored.   (50)
178 PORTRAITS—Types of Beauty, etc., after MIDDLETON, RAE-
BURN and others.   (18)
179 PORTRAITS of Illustrious Americans—GENL. SCOTT, J. J.
AUDUBON, and others.   Proofs on INDIA PAPER.   (5)

180 PORTRAITS. A similar lot. (5)

181 PORTRAITS. A similar lot. (6)

182 ENGRAVINGS, Woodcuts, etc. *Some scarce and curious.*

183 AMERICAN VIEWS—Battle Scenes, Manufactories, etc. (52)

184 NEW YORK CITY and Vicinity illustrated by Engravings, Woodcuts, Photographs, etc. (174)
This collection of nearly 200 pieces is admirably adapted for illustrating a history of the city from its earliest times.

185 HOGARTH (W.). 'Fac-similes of the Original Plates illustrating "Hudibras," "The Four Stages of Cruelty," etc. (14)

186 MARSHALL (W. E.). Portrait of REV. H. W. BEECHER. Fine impression (slightly damaged). (2 duplicates)

187 ORIGINAL SKETCHES in Pencil, Ink and Crayons, and 4 Designs for Frescoes, etc. (8)

188 ETCHINGS—Portraits, etc. Proofs (some in unfinished state). (5)

189 PORTRAITS of French Statesmen, Authors, etc. (12)

190 ETCHINGS—Various French. (25)
Some of these are particularly curious on account of the utter disregard of perspective in their composition.

191 ETCHINGS—Various. A similar lot. (25)

192 ETCHINGS—Various, representing Flowers, Fruit, Foliage, etc. (12)

193 ETCHINGS—Paris and its Environs. (15)

194 ETCHINGS. A similar lot. (16)

195 PARIS—during the Bombardment, and Destruction afterwards by the Communists. (18)

196 MISCELLANEOUS ENGRAVINGS. (22)

197 ETCHINGS—Various. Proofs before Letters. (21)

198 MURILLO (Estevan). "Ecce Homo," engraved by H. Röse, PROOF BEFORE LETTERS ON INDIA PAPER, and 7 others. (8)

199 CARRACCI (Annibal), "Christ in the Garden," engraved by HUFFAM, OPEN LETTER PROOF, and 12 others. (13)

200 PORTRAITS of GENL. GRANT, SAML. COLT, and others. Proofs. (6)

201 PORTRAITS—Various. Proofs before Letters. (6)

202 MURILLO (Estevan), "St. John," Proof before Letters, and 8 others. (9)

203 "FEMALE FIGURES"—a pair. PROOFS UPON SATIN. Mounted. (2)

204 "BITCH AND CUBS." PROOF ON SATIN. Mounted.

205 "VASE OF FLOWERS." PROOF ON SATIN. Mounted.

206 MILITARY SCENE. PROOF UPON SATIN. Mounted.

207 ETCHINGS—Various. Proof before Letters. (9)

208 CASANOVA (Antonio). "Un Coin de Jardin." *Engraved by* CHAMPOLLION. Open Letter Proof. Original impression (slightly damaged).

209 CUNNINGHAM. NELL GWYN ILLUSTRATIONS. 11 *fine portraits*, INDIA PROOFS BEFORE LETTERS, *comprising* NELL GWYN, CATHERINE of BRAGANZA, the DUCHESSES of CLEVELAND, RICHMOND, and PORTSMOUTH, the COUNTESS of GRAMMONT [LA BELLE HAMILTON], CHESTERFIELD, DORSET, and SUNDERLAND, LADY DENHAM, etc. (11)

210 CUNNINGHAM. The Same. INDIA PROOFS BEFORE LETTERS. (11)

211 CUNNINGHAM. The Same, but Plain Proofs BEFORE LETTERS. (11)

212 LA FONTAINE. *A series of some* 80 *exquisite engravings by* DUPLESSIS-BERTEAUX, *Proofs before Letters, printed on* INDIA PAPER, and suitable for insertion in any edition of the Tales without inlaying.

213 SCOTLAND. Views, etc., suitable for Book Illustration. (53)

214 MISCELLANEOUS VIEWS, etc., in Europe, Asia and the East. Some on India paper. (50)

215 MISCELLANEOUS VIEWS. A similar lot. (50)

216 MISCELLANEOUS VIEWS. Early View of Mount Vernon, etc. (45)

217 ETCHED EXAMPLES OF PAINTINGS OLD AND NEW, with Notes by JOHN W. MOLLETT, B.A. 20 *etchings by* FLAMENG, UNGER, LOS RIOS, *etc., etc., after* REMBRANDT, COROT, *etc., etc.* Large 4to, cloth gilt, top edge gilt.
London, 1885

218 EVANGELISCH LUTHERISCHER Gebets Schatz. 8vo, roan.
St. Louis, 1865

219 EVENINGS WITH THE POETS, a Collection of Favorite Poems by Famous Authors. *Illustrated by* DORÉ, FOSTER, *etc., and carmine borders.* 4to, cloth, gilt edges. N. Y., 1880

220 FAIR HARVARD. 12mo, cloth. N. Y., 1869
221 FAIRHOLT (F. W.). Homes, Works and Shrines of English Artists, with Specimens of their Styles, to which is added Rambles in Rome. *Illustrated with numerous wood engravings.* 4to, cloth, gilt edges. London, 1873

222 FAIRHOLT. The Same. Another copy. Cloth (soiled).

223 FAMOUS FRENCH AUTHORS.—Biographical Portraits of Distinguished French Writers, by THEOPHILE GAUTIER, EUGENE DE MIRECOURT, etc. *Illustrated.* 12mo, cloth, top edge gilt. N. Y., 1879

224 FAXON (C. E.) *and* EMERTON (J. H.). Beautiful Ferns, with Descriptive Text by DANIEL CADY EATON. *Colored plates from original water-color drawings after nature by* FAXON *and* EMERTON. Large 4to, cloth, gilt edges. Boston, 1882

225 FIELDING (Henry). Works. *Portrait and vignette title.* Large 8vo, cloth, top edge gilt. Brooklyn, *n. d.*

226 FIELDING. Tom Jones. Traduction par WAILLY. 2 vols. 12mo, half morocco (one cover loose). Paris, 1841

227 FIGUIER (Louis). Vegetable World. *With* 473 *illustrations.* Small 8vo, cloth. N. Y., n. d.

228 FISHER (R. S.). Statistical Gazetteer of the United States. Thick 4to, half roan (rubbed). N. Y., 1867

229 FLEETWOOD (John). The Child's Guide. *Numerous steel engravings.* Large 4to, cloth, gilt (back cover damaged). N. Y., 1878

230 FLETCHER (Rev. Alex., *D. D.*). Scripture Natural History. *With* 265 *illustrations.* 2 vols. thick small 4to, new half morocco, gilt. N. Y., n. d.

231 FORD (W. C.). American Citizen's Manual. 2 vols. square small 8vo, cloth. N. Y., 1882–83

232 FROTHINGHAM (O. B.). Cradle of the Christ. Small 8vo, fresh cloth. N. Y., 1877

233 FROUDE (J. A.). CÆSAR, a Sketch. *Portrait.* Small 8vo, cloth. N. Y., n. d.

234 FULLER (Rev. Andrew). Works, 8 vols.; RICHARD CUM-
BERLAND'S Dramatic Memoirs; SCOTT'S Divine Trinity.
Together 10 vols. 8vo, sheep.

235 GARFIELD (J. A.). Life and Work of, by J. C. RED-
PATH. *Profusely illustrated.* Thick 8vo, cloth.
Cincinnati, 1881

236 GASKELL (Prof. G. A.). Compendium of Forms, Educa-
tional, Social, Legal and Commercial. *Illustrated.* 4to,
mottled sheep, marbled edges. Chicago, 1883

237 GASPARIN (Agenor de), by BOREL, translated by HOWARD.
Small 8vo, cloth. N. Y., 1881

238 GATES (E. M. H.). Your Mission. *Illustrated by* CHURCH,
HARPER *and* ALEXANDER. 4to, cloth, gilt edges
N. Y., 1882

239 GELL (Sir Wm.) *and* GANDY (J. P.). Pompeii, Its Destruction and Re-Discovery; with Descriptions of the Arts and
Architecture of its Inhabitants. *Numerous steel engravings.* Large 4to, cloth, gilt edges (cover loose).
N. Y. [London], n. d.

240 GENIN (T. H.). Selections from Writings of, with Life. 2
portraits. Large 8vo, cloth. N. Y., 1869

241 GERMANY AND SWITZERLAND. Sketches of Travel through.
12mo, cloth. Charleston, 1867

242 GERSTAECKER (F.). Journey Round the World. 12mo,
cloth. N. Y., 1855

243 GILLMORE (Parker, " *Ubique* "). Prairie and Forest, Field
Sports of North America. *Illustrations.* Small 8vo,
fresh cloth. London, 1881

244 GILLMORE. A Ride Through Hostile Africa. *With illustrations by* ELWES. 8vo, new cloth, uncut. London, 1881

245 GOETHE. Letters from Switzerland and Travels in Italy;
Dramatic Works, including IPHIGENIA, TASSO, etc.;
Faust, a Tragedy, also Clavigo, EGMONT, etc. 3 vols.
small 8vo, cloth, top edges gilt. Boston, 1882–84

246 GOETHE. Autobiography; Elective Affinities, Sorrows of
Werther, Travels in Italy, etc.; Poems, IPHIGENIA, TOR-
QUATO TASSO, GOETZ VON BERLICHINGEN. Together 3
vols. Small 8vo, cloth (not uniform color).
N. Y. and Boston, 1882

247 GOETHE. Faust, Clavigo, EGMONT, and Wayward Lover. Small 8vo, cloth, top edge gilt. Boston, 1884

248 GOETHE. Sorrows of Werther, Elective Affinities, and a Novelette. *Illustrated.* Small 8vo, cloth, top edge gilt. Boston, 1884

249 GOLDEN SANDS. Translated by MACMAHON. *Illustrated by* WENTWORTH. 4to, fresh cloth, gilt. N. Y., 1883

250 GOLDONI (Carlo). Commedie. 9 vols. in 6. Small 8vo, boards. Venice, 1753–57

251 GOLDSMITH (Oliver). History of Earth and Animated Nature. *Illustrated.* 4 vols. 8vo, cloth. N. Y., 1881

252 GOODALE. All Round the Year. *Illustrated.* Small 4to, new cloth (bound upside down). N. Y., 1881

253 GRANGE (The). Study in Science of Society, by "Gracchus Americanus." 12mo, cloth. N. Y., 1874

254 GRAY (G. H.). Mystic Circle of Masonry. *Cuts.* Small 8vo, cloth. Cincinnati, 1867

255 GREAT AMERICAN SCULPTURES by WILLIAM J. CLARK, Jr. *With* 12 *superb steel engravings,* INDIA PROOFS. Large 4to, cloth gilt, edges gilt. Phila., n. d.

256 GREATOREX (Eliza). Summer Etchings in Colorado, with Introduction by GRACE GREENWOOD. *Numerous illustrations.* 4to, cloth, gilt edges. N. Y., 1873

257 GREELEY (Horace). Autobiography. *Illustrated.* Thick 8vo, cloth. N. Y., 1872

258 GREENE (Nathaniel, *Major-General*). Life of, by G. W. GREENE. *Portrait.* Vol. I. Large 8vo, cloth, uncut. N. Y., 1867

259 GRIFFITH (J. W.) *and* HENFREY (Arthur). Micrographic Dictionary: a Guide to the Examination and Investigation of Structure and Nature of Microscopic Objects. *Illustrated by* 45 *plates, some colored, and* 812 *woodcuts.* Thick square 8vo, cloth, uncut. London, 1860

260 GRISWOLD (Rufus W.). Poets and Poetry of England in the Nineteenth Century, with Additions by R. H. STODDARD. *Steel portraits.* Large 8vo, cloth, gilt edges. N. Y., n. d.

261 GROTE (George). History of Greece. 4 vols. small 4to, cloth. N. Y., 1881

262 GUILLEMIN (Amédée). Forces of Nature : a Popular Introduction to the Study of Physical Phenomena, translated from the French by Mrs. NORMAN LOCKYER, and edited, with Additions and Notes, by J. NORMAN LOCKYER, F.R.S. *Illustrated by* 11 *colored plates and* 455 *woodcuts.* Thick large 8vo, cloth, top edge gilt.
N. Y., 1872

263 GUIZOT (F.). Popular History of France from the Earliest Times, translated by ROBERT BLACK, M.A. *With* 300 *illustrations by* A. DE NEUVILLE. 6 vols. royal 8vo, new half blue morocco gilt, top edges gilt, others uncut, by MANSELL, successor to HAYDAY. London, 1872

264 GUNNING (W. D.). Life History of our Planet. *Illustrated.* Small 8vo, cloth. N. Y., 1881

265 GUYOT (A.). Earth and Man. *Map and cuts.* 12mo, cloth. Boston, 1861

266 HAKE (A. Egmont). The Story of CHINESE GORDON. *Portraits, maps, etc.* 4to, cloth. N. Y., 1884

267 HAKE. Another copy of the same.

268 HALL (S. C.). A Book of Memories of Great Men and Women of the Age. *Illustrated.* 4to, cloth, gilt edges.
London, *n. d.*

LARGE PAPER COPY OF HALLAM'S WORKS.

269 HALLAM (Henry). Works, *i. e.*:—Middle Ages, 3 vols.; Constitutional History, 3 vols.; Literature of Europe, 4 vols. Together 10 vols. 8vo, cloth, uncut.
Boston, *William Veazie*, 1865–66
LARGE PAPER. Limited edition of one hundred copies.

270 HALLAM. Constitutional History of England. 3 vols. small 8vo, cloth. N. Y., 1877

271 HALLAM. State of Europe during Middle Ages. 3 vols. small 8vo, cloth. N. Y., *Widdleton*, 1874

272 HARPER's Young People for 1884. *Profusely illustrated.* Thick large 4to, cloth. N. Y., 1884

273 HARPER's New Monthly Magazine. *Profusely illustrated.* 28 parts, large 8vo, sewed. N. Y., 1866–83

274 HARRISON (Prof. James A.). Story of Greece. *Illustrated.* Small 8vo, new cloth. N. Y., 1885

275 Harrison's Spain ; Gilman's History of the American
People; Mackenzie's Switzerland. *Illustrated.* 3 vols.
small 8vo, uniform cloth. Boston, n. d.

276 Haweis (H. R.). American Humorists [*i. e.*, Irving,
Holmes, Lowell, Artemus Ward, Mark Twain and
Bret Harte]. Small 8vo, cloth, uncut. London, 1883

277 Hawthorne (N.). Marble Faun, 2 vols.; and Blithedale
Romance, 1 vol. Together 3 vols. Small 8vo, cloth.
Boston, 1852-60
Original editions and scarce.

278 Hayden (F. V.). Geological and Geographical Atlas of
Colorado and Adjacent Territory. *Colored maps.* Large
folio, half morocco, cloth sides. Washington, 1877

279 HECK (J. G.). ICONOGRAPHIC ENCYCLOPÆDIA
of Science, Literature and Art Systematically Ar-
ranged. Translated from the German with Additions,
and edited by Spencer F. Baird. Vols. 1 and 2 of text
large 8vo, half morocco, rubbed, *and the 500 steel plates
containing upwards of* 12,000 *engravings in* 2 *vols.*, of
which 1 vol. is half morocco (rubbed), and the other full
calf (cover loose), oblong large 8vo. Together 4 vols.
Leipzig, 1844-49, and N. Y., 1851

280 Helps (Sir Arthur). Brevia, Short Essays and Aphorisms.
Small 8vo, cloth, uncut. London, *Chiswick Press*, 1871

281 Hemans (Felicia). Poetical Works. *Red line borders and
illustrated.* Small 8vo, half russia, gilt. N. Y., 1881

282 Hemans. The Same Edition. Half russia (binding damaged).

283 Henry (Joseph). Memorial of. *Portrait.* Thick large
8vo, cloth. (2 copies.) Washington, 1880

284 Herbert (George). The Temple—fac-simile of the First
edition, 1633, with Essay by Shorthouse. Small 8vo,
new boards. London, *Gresham Press*, 1882

285 Herodotus *and* Plutarch, edited for Boys and Girls by
John S. White, LL.D. *With numerous illustrations and
maps.* 2 vols. 4to, new cloth gilt. N. Y., 1884

286 Herrick (S. B.). Wonders of Plant Life under the Mi-
croscope. *Illustrations.* Small 4to, cloth. N. Y., 1883

287 HERVEY (A. B.). Flowers of the Field and Forest, with
Descriptive Text by Rev. A. B. Hervey, with Extracts
from Longfellow, Lowell, Bryant, Emerson and
others. *Illustrated with chromo-lithographs from original
water-color drawings after nature by* Isaac Sprague.
Large 4to, cloth gilt, bevelled sides, gilt edges.
Boston, 1883

287* Hervey. Another copy of the same. (Some pp. loose.)

288 Hints for Home Reading. Small 8vo, cloth. N. Y., 1880
With "Suggestions for Libraries," by G. P. Putnam.

289 HOLBEIN (Hans). ILLUSTRIOUS PERSONS of the
COURT of HENRY VIII., with Biographical Notices
by Edmund Lodge, published by John Chamberlaine,
F.S.A., Keeper of the King's Drawings and Medals.
Fac-similes of original drawings, by Hans Holbein,
*for the portraits of illustrious Persons of the Court
of* Henry VIII., *engraved by* Francis Bartolozzi.
Thick large 4to, new half morocco gilt, edges gilt.
London, 1884

290 Holbein and His Time, by Dr. Alfred Woltmann, translated by F. E. Bunnett. *With* 60 *illustrations.* Square
thick large 8vo, cloth gilt, edges gilt.   London, 1872

291 Holmes (Oliver Wendell). Life of, by E. E. Brown.
Small 8vo, cloth.   Boston, 1884

292 Holy Bible according to the Douay and Rheimish Ver-
sions, with Annotations by Challoner. *Steel plates.*
Thick large 4to, morocco, embossed and gilt, one clasp
(one clasp gone and binding rubbed).   N. Y., 1872

293 Holy Bible with the Apocrypha. *Steel plates.* Thick
large 4to, roan, gilt, clasp (binding rubbed). Phila., 1861

294 Holy Bible. Thick 4to, morocco gilt (rubbed).
N. Y., 1860

295 Holy Bible [pica 4to reference]. Thick large 4to,
morocco, gilt edges (rubbed).   N. Y., 1859

296 Household Book of Poetry, collected and edited by
Charles A. Dana. *With illustrations.* Thick 8vo, roan,
gilt edges (rubbed).   N. Y., 1878

297 Houssaye (Arsene). Les Grandes Dames. 5 vols. in 3.
Square minimo, half morocco. Erotic.   Paris, 1871

298 Howe (J. W.). Winter Home for Invalids. 12mo, cloth.
N. Y., 1875

299 Howitt (William). Popular History of Priestcraft. Small
8vo, half calf (MS. notes). London, 1834
300 Hubbell (J. H.). Legal Directory, 1878-79. Thick 8vo,
sheep. N. Y., 1878
301 Huc (Abbe). Journey through the Chinese Empire. *Map.*
2 vols. 8vo, cloth. N. Y., 1871
302 Hughes (R. W.). Currency Question. 12mo, cloth.
N. Y., 1879
303 Hughes (W. K.). Piece Goods, Yarn and Woollen Tables.
Square large 8vo, cloth. London, 1875
304 Humboldt (A. von). Cosmos. *Portrait.* 5 vols. 12mo,
cloth. N. Y., 1859
305 Humboldt. Views of Nature. *Colored front. and fac-
simile.* Small 8vo, cloth, uncut. London, 1850
306 HUME *and* SMOLLETT. History of England. 20 vols.
8vo, half sheep, gilt. Basle, 1789-94
307 Hume (David). History of England. 5 vols. small 8vo,
cloth. London, *n. d.*
308 Hume. The Same. 6 vols. 12mo, cloth. N. Y., 1870-73
309 Hyneman (Leon). History of Freemasonry in England.
Small 8vo, cloth. N. Y., 1878

310 ILLUSTRATED LONDON NEWS. Vols. 19 [June,
1851—Jan., 1852], 21 and 22 [July, 1852—June, 1853].
Together 3 vols. Folio, half leather (not uniform).
311 Illustrated London News. Vol. 51 [July-Dec., 1867].
Folio, half roan.
312 Illustrated London News. 135 Nos. Folio, paper.
London, 1858-84
313 Ingram (John). Flora Symbolica, or Language and Sentiment of Flowers. *Colored plates, with gilt borders.*
Small 4to, cloth gilt, edges gilt. London, *n. d.*
314 IRVING (Washington). Life and Voyages of Christopher Columbus. *Portraits and views.* 3 vols. 12mo,
cloth, red edges (binding soiled). N. Y., 1861-64
315 Irving. Life and Voyages of Christopher Columbus.
Portrait, map and other illustrations. 12mo, new cloth.
N. Y., n. d.
316 Irving. The same.

317 IRVING. Knickerbocker's New York. *Illustrated with plates, portraits, etc.* Thick 4to, fresh cloth.
GEOFFREY CRAYON EDITION. N. Y., n. d.

318 IRVING. Another copy of the same.

319 IRVING. The Life of WASHINGTON and the History of the American Revolution. *With illustrations.* Large 4to, new cloth, gilt. N. Y., 1883

320 IRVING. Little Britain, Spectre Bridegroom and Legend of Sleepy Hollow. *With illustrations, designed by* C. O. MURRAY *and engraved by* J. D. COOPER. Small 4to, new cloth, gilt. N. Y. [London], n. d.

321 IRVING. Astoria, Anecdotes of Enterprise Beyond the Rocky Mountains. 2 vols. 8vo, cloth.
ORIGINAL EDITION. Phila., 1836

321* IRVING. Sketch Book. *Engraved title.* Small 8vo, new cloth, top edge gilt, others uncut. N. Y., *n. d.*

322 IRVING. Wolfert's Roost; Salmagundi. *Plates.* 2 vols. small 8vo, new cloth, gilt tops.
KNICKERBOCKER EDITION. N. Y., *n. d.*

323 IRVING. MAHOMET and his Successors; *also*, Life and Voyages of CHRISTOPHER COLUMBUS. Together 2 vols. Small 8vo, cloth. N. Y., 1884

324 IRVING. Conquest of Granada. *Portrait and front.* 12mo, cloth. N. Y., 1869

325 JAMES (G. P. R.). One in Thousand; Robber; Ancient Régime; RICHELIEU; PHILIP AUGUSTUS; Gipsy and Gentleman of Old School. *Illustrated.* Thick large 8vo, cloth. N. Y., n. d.

326 JARVES (J. J.). Italian Rambles. Small 4to, new cloth.
N. Y., 1883

327 JEWITT (Llewellynn). Life of WILLIAM HUTTON and History of the HUTTON Family. *With portrait.* Small 8vo, half calf, gilt. London, *n. d.*

328 JEWITT. Haddon Hall. *With upwards of 50 illustrations.* 4to, cloth. London, n. d.

329 JEWITT *and* HALL. Stately Homes of England. *With 380 wood engravings.* Thick 4to, cloth, gilt edges.
N. Y., n. d.

330 JOHNSON (Andrew). Trial of. 3 vols. in 2. 8vo, cloth.
Washington, 1868

331 [JOHNSON (Charles).] Chrysal, or the Adventures of a
Guinea. 4 vols. 12mo, tree calf, gilt.   London, 1783
"A masterly, but caustic satire."—LOWNDES.

332 JOHNSON (Samuel). Lives of Most Eminent English Poets.
2 vols. small 8vo, cloth.   Boston, n. d.

333 KENNAN (George). Tent Life in Siberia. 12mo, new
cloth.   N. Y., n. d.

334 KENT (James). Commentaries on American Law. 4 vols.
8vo, old sheep (broken).   Boston, 1858

335 KLODEN (J. F. von). The Self-Made Man, an Autobiography edited by MAX JAHN. *Portrait.* 2 vols. 8vo,
cloth, uncut.   London, 1876

336 KNICKERBOCKER NOVELS, *i. e.*:—GREEN'S Leaven-
worth Case and Strange Disappearance; NOBLE'S Eunice
Lathrop and Uncle Jack's Executors; Man's a Man for
A' That; FLEMMING'S Cupid and Sphinx; KENNEY'S
Gypsie; BELLAMY'S Breton Mills; ELLIOT'S Bassett Claim;
STODDARD'S Heart of It; LANZA'S Mr. Perkin's Daughter;
KIP'S Nestlenook. Together 12 vols. Square 12mo, fresh
red cloth.   N. Y., 1880–85

337 KNIGHT (Charles). Popular History of England. *Profusely illustrated.* 8 vols. small 8vo, cloth, top edges gilt
(one cover damaged).   N. Y., 1881

338 KNIGHT. The Same, on thinner paper. *Illustrated.* 8
vols. in 4. Small 8vo, cloth.   N. Y., 1880

339 KNIGHT. Half Hours with the Best Authors, *steel portraits*,
2 vols; *also*, Half Hours of English History. Together
3 vols. 8vo, uniform cloth gilt.   London, n. d.

340 KOCK (Paul de). Madeleine. *Front.* 2 vols. in 1. 8vo,
half sheep.   Paris, 1845

341 KORNER (Theodor). Sammtliche Werke. 4 vols. small
8vo, half calf, gilt.   Carlsruhe, 1823

342 LAMB (Charles). ESSAYS OF ELIA. TEMPLE EDI-
TION. *Illustrated by* R. SWAIN GIFFORD, JAMES D.
SMILLIE, CHARLES A. PLATT, F. S. CHURCH. 4to, new
cloth, top edge gilt, others uncut.   N. Y., 1884

343 LAMB. Another copy of the same edition.

344 LAMB. Works. *Portrait.* 5 vols. in 3. Small 8vo, cloth.
N. Y., 1882

345 LAMB. Works. *Portrait.* 5 vols. 12mo, cloth. N. Y., 1878

346 LAMB. Eliana. Small 8vo, cloth. N. Y., 1866

347 LANZA (Clara). A Righteous Apostate. Small 8vo, cloth.
N. Y., 1883

348 LAVATER (J. C.). Essays on Physiognomy. *Profusely illustrated.* Thick 8vo, cloth. N. Y., n. d.

348* LAVATER. Another copy of the same.

349 LEGGO (Wm.). Administration of EARL OF DUFFERIN, late Governor-General of Canada. *Two portraits.* Thick 8vo, cloth. Montreal, 1878

350 LEIFCHILD (J. R.). Higher Mystery of Nature. Small 8vo, fresh cloth. N. Y., 1872

351 LE SAGE. Adventures of Gil Blas. *With 500 illustrations by* GIGOUX. Thick 8vo, cloth. N. Y., 1871

352 LESTER (C. Edwards). America's Advancement: The Progress of the U. S. during their First Century. *With 100 superb steel engravings embellishing scenery, history, biography, statesmanship, literature, science and art.* Thick 4to, half morocco, cloth sides, gilt edges.
N. Y., 1876

353 LESTER. Our First Hundred Years. Thick large 8vo, half roan (binding damaged). N. Y., 1877

354 LESTER. The same. Cloth. N. Y., 1876

355 L'ESTRANGE (Rev. A. G.). History of English Humour, with Introduction upon Ancient Humour. 2 vols. in 1. Thick small 8vo, new cloth, gilt edges. London, n. d.

356 LEVER (Charles). Confessions of Harry Lorrequer. *With numerous illustrations by* "PHIZ." 8vo, cloth, uncut. London, n. d.

357 LEVER. Roland Cashel. *With illustrations by* "PHIZ." 2 vols. 8vo, cloth, uncut. London, 1860

358 LEVER. Knight of Gwynne. *With* "PHIZ" *illustrations.* 2 vols. 8vo, cloth, uncut. London, 1860

359 LEVER. Tony Butler. 3 vols. small 8vo, cloth.
London, 1872

360 LEVER's Roland Cashel; Fritz; Boston Almanacks; Poems of Faith; etc. Together 30 vols. Cloth.

361 LIBRARY OF FAMOUS FICTION. 4 vols. 8vo, cloth.

N. Y., 1880

Includes novels by Feuillet, Yates, Hadermann, Tom Hood, Reade, Braddon, Werner, Black, etc.

362 LIEFDE (John de). Charities of Europe. *Illustrated.* 2 vols. small 8vo, cloth, uncut. London, 1865

363 LINDERMAN (H. R.). Money and Legal Tender in U. S. 12mo, new cloth. N. Y., 1879

364 LITTELL'S LIVING AGE, Oct. 1, 1876, to July 1, 1880, inclusive. Together 15 vols., 8vo, of which 10 vols. are in half calf gilt and 5 vols. are unbound in Nos.

Boston, 1876–80

365 LIVY. Legendary History of Rome, translated from the Original Text of TITUS LIVIUS by GEORGE BAKER. *Illustrated with full-page and smaller engravings.* Large 4to, cloth, gilt edges. N. Y., 1883

366 LOCKWOOD (Henry C.). Abolition of the Presidency. 8vo, cloth. N. Y., 1884

367 LOCKWOOD (M. S.). Handbook of Ceramic Art. *Front.* Small 4to, new cloth. N. Y., 1878

368 LONDON QUARTERLY REVIEW. 34 parts. Large 8vo, sewed. N. Y., 1863–74

369 LONGFELLOW PORTFOLIO, *being a selection of* 75 ARTIST PROOFS *from the original woodcuts, illustrating the new subscription edition of* LONGFELLOW'S *Poetical Works.* Large 4to, in cloth flap portfolio (binding damaged). Boston, *Riverside Press*, n. d.
Edition-de-luxe. No. 488 of limited and numbered edition of 500 sets.

370 LOSSING (Benson J.). Home of Washington, or Mount Vernon and its Associations. *Illustrated.* 4to, cloth.
N. Y., n. d.

371 LOUDON (J. C.). Encyclopædia of Cottage, Farm and Villa Architecture and Furniture. *With more than* 2,000 *illustrations.* Thick large 8vo, half leather.
N. Y. [London], 1883

372 LOUDON. The Same. Half roan (binding damaged).
London, n. d.

373 LOUDON. The Same. Half roan (binding damaged).
London, n. d.

374 Lukin (Rev. J.). Young Mechanic, Boy Engineers, and
Amongst Machines. *Profusely illustrated.* 3 vols. square
small 8vo, cloth, gilt. N. Y., 1884, *etc.*

375 Lynch (W. F.). Dead Sea and Jordan. 12mo, cloth.
Phila., 1852

376 LYTTON (Sir E. Bulwer, *Lord*). Novels. *Fronts. (some
foxed and stained).* 18 vols. small 8vo, half calf gilt.
London, 1853–64
Including Rienzi; Caxtons; Night and Morning; Godolphin;
My Novel, 2 vols.; Last of the Barons; Alice, or the Mysteries;
Lucretia; Ernest Maltravers; Pelham; Disowned; Harold; Zanoni;
Devereux; Pilgrims of the Rhine; Last Days of Pompeii; and
Paul Clifford.

377 LYTTON. Works, *i. e.*:—Caxtons and Leila ; Last Days
of Pompeii and Harold ; Paul Clifford and Eugene
Aram ; Kenelm Chillingley and Rienzi; Pelham and
Lucretia; Last of Barons and Pausanias; Devereux and
Disowned; Ernest Maltravers and Alice; What will he
Do with It?; Strange Story and Zanoni; Night and Morn-
ing and Godolphin; Parisians and Pilgrims of Rhine.
Fronts. 24 vols. in 12. Small 8vo, cloth. N. Y., 1884

378 LYTTON. Leila, or the Siege of Granada. *With beautiful steel engravings executed under superintendence of*
Charles Heath. 8vo, cloth, gilt edges. London, 1850

379 Lytton (Robert, *Lord*). Poetical Works of "Owen
Meredith." *Red line borders and portrait.* Small 8vo,
morocco, gilt edges. N. Y., n. d.

380 Lytton. Lucile, by "Owen Meredith." *Portrait, illustrations and carmine borders.* Thick 4to, cloth gilt, edges
gilt. N. Y., 1884

381 MACAULAY (T. B., *Lord*). History of England. *Portrait.* 5 vols. in 2. Thick large 8vo, cloth (binding
damaged). Phila., 1881

382 Macaulay. The same. *Portrait.* 5 vols. small 8vo, new
cloth. Phila., n. d.

383 Macaulay. The same. 5 vols. small 8vo, cloth.
N. Y., n. d.

384 Macaulay. Critical and Miscellaneous Essays. 5 vols.
small 8vo, cloth. N. Y., 1860–61

385 MACKNIGHT's Literal Translation of all the Apostolical
Epistles, 5 odd vols.; CRABB's Family Encyclopædia ;
STANHOPE SMITH's Natural and Revealed Religion. To-
gether 7 vols. 8vo, full leather.

386 MACLEOD (Norman, *D.D.*). Memoir of, by his brother
DONALD MACLEOD. *Portrait.* 8vo, cloth. N. Y., 1876

387 MACLEOD (X. D.). History of Roman Catholicism in North
America. *Portrait.* 8vo, cloth. N. Y., n. d.

388 MACLEOD. History of Devotion to Blessed Virgin Mary in
North America. *Steel portrait.* 8vo, cloth. N. Y., n. d.

389 MACLISE (Daniel, *R.A.*) Pictures by, with Descriptions
and Biographical Sketch of the Painter by JAMES
DAFFORNE. *Steel engravings by eminent engravers.*
Large 4to, cloth gilt (cover loose). London, *n. d.*

390 MACQUOID (K. S.). South Brittany. *With illustrations.*
Small 8vo, cloth, totally uncut. London, n. d.

391 MAGAZINE OF AMERICAN HISTORY. *Illustrated.* 38 parts
(some duplicates). 4to, sewed. N. Y., 1877–82

392 MAIN (David M.). Treasury of English Sonnets, edited
from the original Sources with Notes and Illustrations.
8vo, cloth, top edge gilt, others uncut. N. Y., 1881

393 MALLOCK (Wm. Hurrell). Is Life Worth Living ?, Social
Equality, and Property and Progress. Together 3 vols.
Small 8vo, uniform fresh cloth. N. Y., 1882–84

394 MALO's Asphalte et Bitumes; BIOT's Geometrie; Spirit of
Missions, 1854 ; Cambridge Mechanics ; DODDRIDGE's
Family Expositor ; Dictionnaire Francois-Allemand ;
BOYLE's Works, Vol. 4. Together 7 vols. Half bound.

395 MARGARET, of Navarre. The Heptameron. 12mo, cloth.
N. Y., n. d.

396 MARTIN (R. Montgomery). The British Colonies; their
History, Extent, Condition and Resources. *Steel portraits and maps.* 6 vols. in 3. Thick large 8vo, half
calf, cloth sides, marbled edges. London, n. d.
Includes British North America, Australia, New Zealand, Tas-
mania, Africa, West Indies, British India, Ceylon, etc.

397 MARTINEAU (Harriet). Illustrations of Political Economy.
3 vols. minimo, half calf (binding damaged).
London, 1832

398 MASON (Geo. C., *architect*). Newport and its Cottages. *With numerous heliotype plates.* Thick large 4to, roan, gilt edges, bevelled sides (some margins inked).

Boston, 1875

399 MAYHEW (Henry). London Labour and London Poor. *Numerous illustrations.* 3 vols. 8vo, cloth (one cover loose). London, *n. d.*

400 MEDBERRY (J. K.). Men and Mysteries of Wall Street. *Illustrated.* 12mo, cloth. N. Y., 1878

MEDICAL AND SURGICAL LITERATURE.

401 ACTON (W.). Functions and Disorders of Reproductive Organs. 8vo, cloth. Phila., 1867

402 ALLEN (Harrison). System of Human Anatomy, including its Medical and Surgical Relations. *With numerous plates, many colored, and woodcuts.* 4 vols. large 4to, sewed and loose in fresh cloth portfolios. Phila., 1882–83
    The four parts comprise:—I. Histology, by E. O. Shakespeare; II. Bones and Joints ; III. Muscles and Fasciæ ; IV. Arteries, Veins and Lymphatics.

403 AMERICAN JOURNAL OF THE MEDICAL SCIENCES. January, 1865, to January, '68, inclusive. Together 13 parts. 8vo, sewed. Phila., 1865–68

404 AMERICAN JOURNAL OF OBSTETRICS. 50 parts. Large 8vo, sewed (some duplicates). N. Y., 1868–82

405 BARTHOLOW (R.). Manual of Hypodermic Medication. *Illustrations.* 12mo, cloth. Phila., 1869

406 BARWELL (R). Diseases of the Joints. *Numerous cuts.* 8vo, new cloth. N. Y., 1881

407 BEASLEY (H.). Book of Prescriptions. 12mo, sheep. Phila., 1855

408 BECK (T. R., *and* J. B.). Elements of Medical Jurisprudence, revised by GILMAN. 2 vols. 8vo, sheep. Phila., 1863

409 BEDFORD (G. S.). Diseases of Women and Children. Thick 8vo, sheep. N. Y., 1867

410 BELL (Benjamin). System of Surgery. *Illustrated.* 6 vols. 8vo, old sheep. Edinburgh, 1787–88

411 BENNET (J. H.). Inflammation of Uterus. Large 8vo, cloth. Phila., 1864

412 BERMINGHAM (E. J.). Encyclopædic Index of Medicine and Surgery. Thick 8vo, fresh sheep. N. Y., 1882

413 BOSTON MEDICAL and Surgical Journal from June 5, 1879, to July 13, 1882, inclusive [*i. e.*, Vols. 101—6 complete]. Together 163 parts. 4to and large 8vo, sewed.
Boston, 1879–82

414 BRAITHWAITE'S Retrospect of Practical Medicine and Surgery. Parts 15 to 17 inclusive ; Parts 20 and 21; Part 38; Part 52 to 62 inclusive; Part 66 and 67; Dec., '80. Together 20 parts. Large 8vo, sewed. N. Y., 1847–80

415 BRANDS (O. M.). Lessons on the Human Body. *Cuts.* Small 8vo, new cloth. (2 copies.) Boston, n. d.

416 BUCK (A. H.). Diagnosis and Treatment of Ear Diseases. *Cuts.* 8vo, fresh cloth. N. Y., 1880

417 BUCKNILL *and* TUKE. Manual of Psychological Medicine and Insanity. *Front.* Large 8vo, cloth. Phila., 1858

418 BUMSTEAD (F. J.). Pathology and Treatment of Venereal Diseases. *Illustrated.* Large 8vo, cloth. Phila., 1864

419 BYFORD (W. H.). Theory and Practice of Obstetrics. *Cuts.* 8vo, cloth. N. Y., 1873

420 BYFORD. Inflammation and Displacements of Unimpregnated Uterus. 8vo, cloth. Phila., 1864

421 CARPENTER (W. B.). Principles of Human Physiology, with Additions by F. GURNEY SMITH. *With* 261 *cuts.* Large 8vo, sheep. Phila., 1862

422 CARPENTER'S Human Physiology, by CLYMER; TURNER'S Chemistry. Together 2 vols. Sheep.

423 CARSON (Joseph). Materia Medica and Pharmacy Lectures. Large 8vo, cloth. Phila., 1855

424 CATHELL (D. W.). Physician Himself. 8vo, cloth, red edges. Baltimore, 1883

425 CAZEAUX (P.). Midwifery, Pregnancy and Parturition, edited by BULLOCK. *With* 140 *illustrations.* Large 8vo, sheep. Phila., 1860

426 CHAMBERS (T. K.). Renewal of Life—Lectures, Chiefly Clinical. Large 8vo, cloth. Phila., 1866

427 CHARCOT (J. M.). Clinical Lectures on Diseases of Old Age, translated by HUNT, with Lectures by LOOMIS. 8vo, fresh cloth. N. Y., 1881

428 CHEYNE (G.). Gout. Small 8vo, old sheep. London, 1722

429 CHILDBIRTH. Small 8vo, cloth. N. Y., 1845

430 CHURCHILL (F.). Diseases of Women, with Notes and Additions by CONDIE. Large 8vo, cloth.   Phila., 1857

431 CLARKE (W. F.). Manual of Practice of Surgery. · *Cuts.* 8vo, fresh cloth.   N. Y., 1879

432 CONDIE (D. F.). Diseases of Children. Large 8vo, sheep.   Phila., 1858

433 COULSON (W. J.). Diseases of Bladder and Prostate Gland. 8vo, fresh cloth.   N. Y., 1881

434 DALTON (J. C., jr.). Human Physiology. *With* 254 *illustrations.* Large 8vo, sheep.   Phila., 1859

435 DELAFIELD *and* STILLMAN. Manual of Physical Diagnosis. *Illustrated.* 4to, cloth.   N. Y., 1878

436 DICKINSON (W. H.). Treatise on Albuminuria. *Cuts.* 8vo, fresh cloth.   N. Y., 1881

437 DISEASES OF INTESTINES and Peritoneum, by BRISTOWE, WARDELL, BEGBIE, HABERSHON, CURLING and RANSOM. *Cuts.* 8vo, fresh cloth. (2 copies.)   N. Y., 1879

438 DUNGLISON (R. J.). Practitioner's Reference Book. *Cuts.* 8vo, cloth.   Phila., 1878

439 DUNGLISON (Robley). New Remedies, with Formulæ. Thick large 8vo, cloth.   Phila., 1856

440 DUNGLISON. Dictionary of Medical Science. Thick large 8vo, sheep.   Phila., 1860

441 ELLIS (B.). Medical Formulary, edited by MORTON. 8vo, cloth.   Phila., 1849

442 ELLIS (Edward). Diseases of Children, with a Formulary. 8vo, fresh cloth.   N. Y., 1879

443 ELLIS (G. V.) *and* FORD (G. H.). Illustrations of Dissection. *Colored plates.* 2 vols. 8vo, fresh cloth.   N. Y., 1882

444 ERICHSEN (J.). Science and Art of Surgery. *With* 417 *woodcuts.* Thick large 8vo, sheep.   Phila., 1860

445 ETHICAL SYMPOSIUM, Medical Ethics and Etiquette. Small 8vo, new cloth.   N. Y., 1883

446 FLINT (Austin). Principles and Practice of Medicine. Thick large 8vo, sheep.   Phila., 1868

447 FORBES, TWEEDIE *and* CONOLLY. Cyclopædia of Practical Medicine, revised, with additions by ROBLEY DUNGLISON. 4 vols. large 8vo, sheep.   Phila., 1867

448 FORDYCE (W.). Venereal Disease. Small 8vo, sheep.   London, 1777

449 FOTHERGILL (J. M.). The Maintenance of Health. Small 8vo, half cloth. N. Y., n. d.

450 FOWNES' Chemistry for Students; DRAPER's Chemistry; CUTTER's Anatomy and Physiology; etc. Together 5 vols. Full sheep.

451 FRERICHS (F. T.) Clinical Treatise on Diseases of the Liver. *Cuts.* 3 vols. 8vo, fresh cloth. N. Y., 1879

452 GARDNER's Conjugal Sins ; N. Y. Medical Society Transactions ; etc. Together 9 pieces. Sewed.

453 GARRATT (Alfred C.). Myths in Medicine and Old-Time Doctors. Small 8vo, fresh cloth. N. Y., 1884

454 GILBERT (Luther M.). Home Physician. Small 4to, cloth. N. Y., 1883

455 GRIESINGER (W.). Mental Pathology and Therapeutics, translated by ROBERTSON and RUTHERFORD. 8vo, fresh cloth. N. Y., 1882

456 GUTTMANN (Paul). Handbook of Physical Diagnosis, translated by NAPIER. *With colored plate and* 89 *woodcuts.* 8vo, fresh cloth. N. Y., 1880

457 HARTSHORNE (H.). Principles and Practice of Medicine. 12mo, cloth. Phila., 1871

458 HEMMING (W. D.). Forensic Medicine and Toxicology. Small 8vo, cloth. London, 1878

459 HENOCH (Edward). Lectures on Diseases of Children. 8vo, fresh cloth. N. Y., 1882

460 HILL (Berkeley). Essentials of Bandaging. *Cuts.* Small 8vo, cloth. London, 1876

461 HILTON (John). On Rest and Pain, edited by JACOBSON. *Cuts.* 8vo, fresh cloth. N. Y., 1879

462 HODGE (Hugh L.). Principles and Practice of Obstetrics. *Numerous lithographic plates and woodcuts.* Large 4to, cloth. Phila., 1864

463 HUNTER *and* RICORD. Venereal Disease, translated and edited by BUMSTEAD. Large 8vo, cloth. Phila., 1859

464 JOHNSON (Lawrence). Medical Formulary. 8vo, fresh cloth. N. Y., 1881

465 KANE (H. H.). Opium Smoking in America and China. Small 4to, cloth. N. Y., 1882

466 KELSEY (C. B.). Diseases of Rectum and Anus. *Cuts.* 8vo, fresh cloth. N. Y., 1882

467 Keyes (E. L.). Venereal Diseases, including Stricture of
Male Urethra. *Cuts.* 8vo, fresh cloth.   N. Y., 1880
468 Kirke (W. S.). Manual of Physiology. *Cuts.* 12mo,
sheep.   Phila., 1857
469 Lambert's Pictorial Anatomical Charts. Plates 1 to 6
inclusive ; varnished, canvas mounted and on rollers.
(6 pieces.)
470 Laurence (W.). Treatise on Diseases of the Eye, edited
by Isaac Hays. *With* 243 *illustrations.* Thick large 8vo,
sheep.   Phila., 1854
471 Liveing (R.). Treatment of Skin Diseases. Small 8vo,
cloth, uncut.   N. Y., 1878
472 Longstreth (M.). Rheumatism, Gout and Allied Disorders. 8vo, fresh cloth.   N. Y., 1882
473 Loomis (A. L.). Lectures on Fevers. Large 8vo, sheep.
N. Y., 1877
474 Lyman (H. M.). Artificial Anæsthesia and Anæsthetics.
Cuts. 8vo, fresh cloth.   N. Y., 1881
475 Mackenzie (Morell). Diseases of Pharynx, Larynx and
Trachea. *Cuts.* 8vo, fresh cloth.   N. Y., 1880
476 Mann (M. D.). Prescription Writing. Minimo, cloth.
N. Y., 1878
477 Mauriceau (A. M.). Married Woman's Companion.
Minimo, cloth.   N. Y., 1855
478 Maygrier (J. P.). Midwifery Illustrated. 82 *plates.*
Large 8vo, cloth (back broken).   N. Y., 1836
479 MEDICAL and ANTHROPOLOGICAL STATISTICS
of Provost-Marshal General's Bureau derived from Records of over a Million Recruited, Drafted Men, Substi-
tutes and Enrolled Men. Compiled by J. H. Baxter,
M.D. *Maps and diagrams.* 2 vols. thick large 4to, cloth.
Washington, 1875
480 Medical Examiner. Vol. 1. Large 8vo, half roan (no
title).   Phila., 1838
481 Medical News and Library. Jan., 1865, to Nov., '67, inclusive [August, '65, short]. Together 36 parts. Large
8vo, sewed.   Phila., 1865–67
482 MEDICAL RECORD. Jan., 1873, to Dec., 1875, inclusive.
Together 99 parts [Vols. 8 and 10 complete; Vol. 9 lacks
a No.]. 4to, sewed.   N. Y., 1873–75

483 MEDICAL RECORD. Vols. 15 to 20 inclusive [Jan. 4, 1879—Dec. 31, 1881]. Together 6 vols. 4to, half sheep, cloth sides. N. Y., 1879–81

484 MEDICAL RECORD. For year 1882 [one No. short]. Together 51 parts. 4to, sewed. N. Y., 1882

485 MEDICAL AND SURGICAL HISTORY OF THE WAR of the Rebellion 1861–65. 4 vols. [*i. e.*, Part 1, Vol. 1, Medical History and Appendix ; Part 1, Vol. 2, Surgical ᐧ History; Part 2, Vol. 1, Medical History ; Part 2, Vol. 2, Surgical History]. *Profusely illustrated with plates (some tinted) and cuts.* Thick large 4to, cloth. Washington, 1875–79

486 MEDICAL AND SURGICAL HISTORY OF THE WAR. Part 3, Vol. 2, Surgical History. *Plates, some colored, cuts, etc.* Thick large 4to, cloth. Washington, 1883

487 MEDICAL AND SURGICAL HISTORY OF THE WAR. Part 1, Vol. 2, Medical History. Thick large 4to, cloth. Washington, 1879

488 MEIGS *and* PEPPER. Diseases of Children. Thick large 8vo, sheep. Phila., 1870

489 MUNDE (Paul F.). Minor Surgical Gynecology. *With* 300 *illustrations.* 8vo, fresh cloth (a few pp. soiled). N. Y., 1880

490 NAPHEYS (G. H.). Modern Surgical Therapeutics. Thick 8vo, cloth. Phila., 1879

491 NAPHEYS. Modern Therapeutics. Thick 8vo, cloth. Phila., 1877

492 NAPHEYS. The Masculine Function. 12mo, cloth. Phila., 1871

493 NELSON (J. H.). Druggist's Hand-Book of Private Formulas. 12mo, cloth. N. Y., 1882

494 NEW ENGLAND MEDICAL MONTHLY. 17 parts. 4to, sewed. Bridgeport, 1883–85

495 NEW YORK MEDICAL SOCIETY TRANSACTIONS for Years 1807–31, '58 ; '61 to '67 inclusive ; '69 to '74 inclusive ; '76 to '84 inclusive. Together 23 vols. 8vo, cloth, some with gilt top edges. Albany, etc., 1858–84

496 N.Y. STATE HOMŒOPATHIC MEDICAL SY. Transactions, 1864–66, also '69 ; N. Y. Medical Sy. Transactions, 1869 ; N. Y. Eclectic Medical Sy. Transactions, 1868–69. Together 6 vols. 8vo, cloth.

497 Noyes (H. D.). Diseases of the Eye. *Cuts.* 8vo, fresh cloth.   N. Y., 1881

498 O'Dea (J. J.). Suicide—its Philosophy, Causes and Prevention. Small 8vo, fresh cloth.   N. Y., 1882

499 Parrish (E.). Treatise on Pharmacy. *With 238 illustrations.* Thick 8vo, cloth.   Phila., 1865

500 Pavy (F. W.). Food and Dietetics. 8vo, fresh cloth.   N. Y., 1881

501 Phillips (C. D. F.). Materia Medica and Therapeutics— Vegetable Kingdom and Inorganic Substances. 2 vols. 8vo, fresh cloth (binding not uniform).   N. Y., 1879–82

502 Piffard (H. G.). Materia Medica and Therapeutics of the Skin. *Cuts.* 8vo, fresh cloth.   N. Y., 1881

503 Poulet (A.). Treatise on Foreign Bodies in Surgical Practice. *Cuts.* 2 vols. 8vo, fresh cloth.   N. Y., 1880

504 Putzel (L.). Common Forms of Functional Nervous Diseases. 8vo, fresh cloth.   N. Y., 1880

505 Ranney (A. L.). Practical Medical Anatomy. *Cuts.* 8vo, fresh cloth.   N. Y., 1882

506 Repertorio Medico (El). *Illustrated.* Vol. 1. 4to, half roan.   N. Y., 1883

507 Rice (C.). Posological Table. *Cuts.* Small 8vo, cloth.   N. Y., 1879

508 Ringer (S.). Handbook of Therapeutics. Small 8vo, cloth.   N. Y., 1871

509 Roosa *and* Ely. Ophthalmic and Otic Memoranda. Minimo, cloth.   N. Y., 1876

510 Rosenthal (M.). Clinical Treatise on Diseases of Nervous System, with Preface by Charcot, translated by Putzel. 2 vols. 8vo, fresh cloth.   N. Y., 1879

511 Routh (C. H. F.). Infant Feeding. 8vo, fresh cloth.   N. Y., 1879

512 Ryan (Michael). Philosophy of Marriage. 12mo, cloth.   Phila., 1870

513 Salter (H. H.). Asthma, its Pathology and Treatment. *Cuts.* 8vo, fresh cloth.   N. Y., 1882

514 SANGER (W. W.). History of Prostitution. Large 8vo, cloth.   N. Y., 1869

515 Savage (Hy.). Surgery, Surgical Pathology and Surgical Anatomy of the Female Pelvic Organs. *Plates and cuts.* 8vo, fresh cloth.   N. Y., 1880

516 SCANZONI (F. W. von). Diseases of Sexual Organs of Woman, annotated by GARDNER. *With 60 illustrations.* Large 8vo, cloth. N. Y., 1861

517 SIMS (J. Marion). Clinical Notes on Uterine Surgery. *Cuts.* 8vo, cloth. N. Y., 1869

518 SLADE (D. D.). Diphtheria. 12mo, cloth. Phila., 1864

519 SMITH (S.). Handbook Surgical Operations. *Cuts.* Small 8vo, cloth. N. Y., 1863

520 STILLE (A.). Therapeutics and Materia Medica. 2 vols. thick large 8vo, sheep. Phila., 1860

521 SURGICAL INSTRUMENT CATALOGUE. *Cuts.* 8vo, cloth. London, 1876

522 SWAYNE (J. G.). Obstetric Aphorisms. *Cuts.* 12mo, cloth. Phila., 1870

523 TAIT (Lawson). Diseases of Women. 8vo, fresh cloth. N. Y., 1879

524 TANNER (T. H.). Index of Diseases and their Treatment. Large 8vo, cloth. Phila., 1867

525 THOMAS (T. Gaillard). Practical Treatise on Diseases of Women. *With 225 illustrations.* Large 8vo, sheep. Phila., 1869

526 THOMPSON (E. S.). Colds and Coughs. Small 8vo, cloth. Phila., 1878

527 THOMPSON (J. H.). Report of Columbia Hospital for Women and Lying-in Asylum, Washington. *Illustrated.* Large 4to, cloth. Washington, 1873

528 TIDY (C. M.). Legal Medicine. *Colored front.* 2 vols. 8vo, fresh cloth. N. Y., 1882
Includes—Monstrosities and Hermaphrodism.

529 TILT (E. J.). Uterine Therapeutics and Diseases of Women. 8vo, fresh cloth. N. Y., 1881

530 TILT. Uterine Therapeutics. Large 8vo, cloth. N. Y., 1869

531 TODD (R. B.). Clinical Lectures on Certain Acute Diseases. Large 8vo, cloth. Phila., 1860

532 TOYNBEE (J.). Diseases of the Ear. *Illustrated.* Large 8vo, cloth. Phila., 1865

533 TROUSSEAU *and* PIDOUX. Treatise on Therapeutics. Ninth edition, by PAUL. Translated by D. F. LINCOLN. 3 vols. 8vo, fresh cloth. N. Y., 1880

534 WEST (C.). Diseases of Infancy and Childhood. Thick large 8vo, cloth. Phila., 1860

535 WHAT TO OBSERVE in Medical Cases. 12mo, cloth.
　　　　　　　　　　　　　　　　　　　　　　Phila., 1859
536 WILSON (Erasmus). Human Anatomy, edited by Go-
　　BRECHT. *With 397 cuts.* Large 8vo, sheep (title soiled).
　　　　　　　　　　　　　　　　　　　　　　Phila., 1858
537 WILSON. Diseases of the Skin. *With plates and illustra-
　　tions.* Thick large 8vo, cloth. 　　　　　Phila., 1865
538 WILSON (J. C.). Treatise on Continued Fevers, with In-
　　troduction by DA COSTA. *Cuts.* 8vo, fresh cloth.
　　　　　　　　　　　　　　　　　　　　　　N. Y., 1881
539 WINSLOW (Forbes). Obscure Diseases of the Brain and
　　Mind. Large 8vo, cloth. 　　　　　　　　Phila., 1866
540 WITTHAUS (R. A.). General Medical Chemistry. 8vo,
　　fresh cloth. 　　　　　　　　　　　　　　N. Y., 1881
541 WOOD (G. B.). Practice of Medicine. 2 vols. thick 8vo,
　　sheep. 　　　　　　　　　　　　　　　　Phila., 1858
542 WOOD. Therapeutics and Pharmacology, or Materia Med-
　　ica. 2 vols. 8vo, sheep. 　　　　　　　　Phila., 1856
543 WOOD *and* BACHE. U. S. Dispensatory. Thick 8vo,
　　sheep. 　　　　　　　　　　　　　　　　Phila., 1865
544 WRIGHT (H. G.). Headaches. 12mo, cloth. Phila., 1867
545 WYMAN (M.). Autumnal Catarrh—Hay Fever. *Maps.*
　　8vo, cloth. 　　　　　　　　　　　　　　N. Y., 1872

MEDICAL AND SURGICAL INSTRUMENTS, Etc.

546 HUMAN SKELETON, almost perfect, and comprising
　　some 100 pieces, *i. e.,* Cranium, Pelvis, Vertebræ, Spinal
　　Column, Femurs, Coccyx, and other Osteological Parts.
547 A. M. DAY'S SPLINTS, a set in good order, including
　　Collar Bone, Arm, Leg, Foot, Ankle, Elbow Splints and
　　Rests, of which there are three sizes, and which make in
　　all 　　　　　　　　　　　　　　　　　　(46 pieces)
　　　This set, which is nearly new, cost seventy-five dollars.
548 VAGINAL SPECULUM, by TIEMANN.
549 THREE-VALVED SPECULUM, by OTTO & REYNDERS.
550 SIMS'S Vaginal Speculum, by TIEMANN.
551 BIVALVE Vaginal Speculum, by TIEMANN.
552 FENESTRATED Rectal, and FERGUSON'S Vaginal Specula
　　　　　　　　　　　　　　　　　　　　　　(2 pieces)
553 PAIR OF OBSTETRICAL FORCEPS, by J. REYNDERS.

554 OBSTETRICAL INSTRUMENTS, in leather case, and consisting of a pair of DENMAN's Obstetrical Forceps ; Graded Oretis and Blunt Hook ; Pair Curved Abortion Forceps ; Straight Dilating and Placenta Forceps ; Vulsillum Forceps. (6 pieces)

555 ROCHET CASE, leather, containing 14 Surgical Instruments. (1 lot)

556 AMPUTATING INSTRUMENTS, by TIEMANN. A set in mahogany case, four Dissecting Knives, two Saws, etc., etc. (1 lot)

557 DENTAL FORCEPS, seven pairs, and two Dental Elevations. (9 pieces)

558 BRASS STOMACH PUMP, and Back Double-action Syringe, in box.

559 SYRINGES, Aural, Nasal, etc. (4 pieces)

560 SPRING SCARIFICATOR, by TIEMANN, in case.

561 LEATHER CORD Single Spring Truss, by REYNDERS. New.

562 OPHTHALMOSCOPIC HEAD MIRROR, and three Ear Specula. (4 pieces)

563 SILVER AND GERMAN-SILVER Uterine Probes, Catheters, etc. (12 pieces)

564 PRESCRIPTION SCALES (5) and Medical Battery (some imperfect). (6 pieces)

565 FEVER THERMOMETER, by SHEPHERD *and* DUDLEY, in case.

566 FLEXIBLE SINGLE STETHOSCOPE.

567 SYRINGES, Spray Distributors, etc.
(A lot of some 25 pieces)

568 MEDICINE CASES, Prescription Slab, etc. (A lot)

569 LARYNGIAL FORCEPS, Exploring Needle, Porte-caustic and Handle and Probes. (6 pieces)

570 UTERINE SCARIFICATOR; Needle Holder; Speculum; Probe, etc. (9 pieces)

571 SPATULA; Dressing Forceps, 2 pairs; Silver Male and Female Catheter ; Aural Speculum ; Diabetonometer ; Sponge Carrier; etc. (9 pieces)

572 ELECTROID INSTRUMENTS—Uterine, Tongue, Rectal, Eye, etc. A set in box.

573 MEW (James). Types from Spanish Story, or The Old Manners and Customs of Castile. *With 36 proof etchings on China paper by* R. De Los Rios. 4to, illuminated cloth, top edge gilt. N. Y., 1884

574 Michelet (Jules). History of France. 2 vols. large 8vo, cloth. N. Y., 1875

575 MILTON (John). Poetical Works, Memoir and Notes— Chandos Edition. *Original illustrations, steel portrait and red line borders.* Thick small 8vo, cloth gilt, edges gilt. London, n. d.

576 Milton. Poetical Works. *Red line border and illustrated.* Small 8vo, roan gilt, edges gilt. N. Y., 1881

576* Milton. The Same Edition. Half russia, gilt.

577 MILTON. Treatise on Christian Doctrine, translated by Sumner. *Fac-similes.* Thick large 4to, morocco, gilt edges (rubbed).
Cambridge and London, *Chas. Knight*, 1825

578 MODERN ETCHINGS OF CELEBRATED PAINT-ERS, with an Essay by John W. Mollett, B.A. 20 *etchings by* Gaillard, Unger, Flameng, *etc., after* Rembrandt, Velasquez, Meissonier, Holbein, *etc., etc.* Large 4to, cloth, top edge gilt. London, 1883

578* Modern Etchings. Another copy of the same.

579 Mongredien (A.). Free Trade Movement in England. Small 8vo, cloth. N. Y., 1881

580 Monk (Maria). Awful Disclosures of the Hotel Dieu Nunnery at Montreal. *Plan.* 12mo, cloth. N. Y., 1836

581 Morrell (Benjamin). Narrative of Four Voyages. *Portrait.* 8vo, calf gilt. N. Y., 1832

582 Morrison (Mary J.). Songs and Rhymes for the Little Ones. Small 8vo, new cloth. N. Y., 1884

583 Morse (S. F. B., *LL.D.*). Life of, by Samuel Irenæus Prime. *Portraits and other illustrations.* Thick 8vo, cloth. N. Y., 1875

584 Morse. Memorial of. *Portrait.* Large 8vo, cloth.
Washington, 1875

585 MULREADY (William, *R.A.*). Pictures, with Descriptions and a Biographical Sketch of the Painter by James Dafforne. 10 *steel engravings after* Mulready *by eminent engravers.* Large 4to, cloth gilt, edges gilt.
London, *n. d.*

586 MURRAY (Rev. John A.). Bible Lyrics. *Profusely illustrated.* 4to, cloth. N. Y., 1870

587 MUSSET (Alfred de). Biography of. Small 8vo, cloth. (2 copies.) Boston, 1877

588 NADAILLAC (Marquis de). Pre-Historic America. Translated by N. D'ANVERS and edited by W. H. DALL. *With* 219 *illustrations.* Thick large 8vo, cloth, top edge gilt, others uncut. N. Y., 1884

589 NATIONAL QUARTERLY REVIEW. 45 parts. Large 8vo, sewed. N. Y., 1862–69

590 NEANDER (A.). Life of CHRIST. Small 8vo, cloth. London, 1871

591 NEVIUS'S Life in China; Eastern Manners and Customs; SAWYER'S Memoir of the REV. STEPHEN SMITH; POPE not Antichrist; ISHAM'S Mud Cabin; etc. Together 10 vols. Cloth.

592 NEWELL (C. M.). KAMEHAMEHA, the Conquering King. *Front.* Small 8vo, cloth. N. Y., 1885

593 NEW TESTAMENT, Revised Version. *With carmine borders.* 8vo, roan, gilt edges. N. Y., 1881

594 NEW TESTAMENT. The Same Edition. 8vo, cloth, red edges.

595 NEW TESTAMENT; BELL'S Correct Language; TODD'S Book of Analysis; Household Reading; Education Commissioner's Report for 1881 (2 copies). Together 6 vols. Cloth.

596 NEWTON (R. Heber). Book of the Beginnings, Right and Wrong Use of Bible, and Womanhood. Together 3 vols. Small 4to, cloth. N. Y., 1881–84

597 N. Y. COMMON COUNCIL MANUALS, 1864–65 (duplicates); N. Y. Legislative Manuals. Together 7 vols.

598 NICHOLS (G. W.). Pottery, How It Is Made, with a full Bibliography of Ceramics. *Illustrated.* Small 8vo, new boards. N. Y., 1878

599 NILES (W.). 500 Majority. Large 8vo, cloth. N. Y., 1872

600 NORCROSS (J.). History of Democracy. 12mo, fresh cloth. N. Y., 1883

601 NORTH AMERICAN REVIEW. 48 parts. Large 8vo, sewed. N. Y., 1879–85

602 North British Review. From Feb., 1863, to April, 1870, inclusive (April, '68, short). Together 28 parts. Large 8vo, sewed. N. Y., 1863–70

603 OPPERT (Ernest). A Forbidden Land—Voyages to the Corea. *With 2 charts and* 21 *illustrations.* 8vo, cloth. N. Y., 1880

604 OUR BRITISH PORTRAIT PAINTERS from Sir Peter Lely to James Sant, with Descriptive and Historical Notices by Edmund Ollier. *With* 16 *steel engravings.* Large 4to, cloth gilt, edges gilt. London, n. d.

605 Our British Portrait Painters. Another copy. (Some plates loose.)

606 Our Poetical Favorites—Selection from Best Minor Poems of English Language. 3 vols. small 8vo, cloth, gilt edges. Boston, 1881

607 PAINE (Thomas). Complete Works. *Portrait.* Thick 8vo, cloth. Chicago, n. d.

608 Pancoast (S.). The Kabbala. *Colored illustrations,* printed in blue type, with red borders. 8vo, cloth. N. Y., 1883

609 Parker (Matthew, *Archbishop*). Correspondence of. 8vo, cloth, uncut. Cambridge, *Parker Society,* 1853

610 Parton (James). Illustrious Men and their Achievements. *Illustrated.* Thick 8vo, cloth, gilt edges. N. Y., n. d.

611 Pen Pictures of Earlier Victorian Authors. Small 4to, fresh cloth, top edge gilt. N. Y., 1884

612 Perrin (Raymond S.). Religion of Philosophy or Unification of Knowledge. Thick 8vo, cloth. N. Y., 1885

613 [Piatt (Mrs. L. K.).] Bell Smith Abroad. *Illustrated.* 12mo, cloth. N. Y., 1855

614 Pictures of Society—Grave and Gay, from the Pens of Popular Authors. *Illustrated by* Millais, Watson, Horsley, Cope, *and numerous other eminent English artists.* 4to, green morocco gilt, bevelled sides, gilt edges. N. Y. [London], 1866

615 Pierrepont (Edward). Fifth Avenue to Alaska. *Maps.* Small 8vo, cloth. N. Y., 1885

616 PLEASONTON (Gen. A. J.). Blue Ray of Sunlight and Color of the Sky. *Front.* 8vo, cloth. Phila., 1877

617 PLUTARCH. Lives, from the Original Greek, with Notes Critical, Historical and Chronological, and a New Life of PLUTARCH translated by JOHN and WILLIAM LANGHORN. 3 vols. large 8vo, half calf gilt, citron edges. London, 1812–13

618 PLUTARCH'S Lives and Histories of HERODOTUS, selected and edited by WHITE. *Illustrated.* Together 4 vols. Square small 8vo, new cloth. N. Y., 1885

619 POPULAR READINGS in Prose and Verse, selected and edited by J. E. CARPENTER. *Portrait.* 4 vols. small 8vo, cloth. London, n. d.

620 POPULAR SCIENCE MONTHLY. *Illustrated.* 180 parts. Large 8vo, sewed (some duplicates). N. Y., 1872–84

621 PRESBYTERIAN CHURCH Throughout the World. Biographical and Historical Sketches. *Portraits and other illustrations.* Thick 8vo, roan gilt (rubbed). N. Y., 1874

622 PRESCOTT (William H.). WORKS AND LIFE, *i. e.*:—Conquest of Mexico, 3 vols.; PHILIP THE SECOND, 3 vols.; FERDINAND and ISABELLA, 3 vols.; Conquest of Peru, 2 vols.; ROBERTSON'S CHARLES THE FIFTH by PRESCOTT, 3 vols.; Life of W. H. PRESCOTT by TICKNOR. *Portraits, etc.* Together 15 vols. Small 8vo, uniform fresh green cloth. Phila., 1882

623 PRESCOTT. History of the Conquest of Mexico. *Fronts. of portraits (one foxed and one stained).* 3 vols. 8vo, cloth, uncut (one cover loose). London, 1844

624 PRESCOTT. Conquest of Peru. *Portraits.* 2 vols. 8vo, cloth, uncut. London, 1847

625 PRESCOTT. Reign of FERDINAND and ISABELLA. *Portraits and fac-simile.* 3 vols. 8vo, cloth, uncut (one cover damaged). London, 1842

626 PRESCOTT. Conquest of Mexico. *Portraits and maps.* 3 vols. large 8vo, cloth, uncut. N. Y., 1847

627 PRESCOTT. FERDINAND and ISABELLA. *Portraits.* 3 vols. 8vo, cloth, uncut (one cover damaged). SCARCE. Boston, 1844

627* PRINCE (L. Bradford). Articles of Confederation *vs.* The Constitution. 12mo, cloth. N. Y., 1867

628 PROCTER (Adelaide A.). Poems, Complete Edition, with Introduction by CHARLES DICKENS. *Portrait, illustrations and carmine borders.* Thick cloth gilt, edges gilt.   N. Y., 1884

629 PROCTER (B. W., "*Barry Cornwall*"). Autobiography and Personal Sketches. *Portrait.* Small 8vo, cloth.
Boston, 1877

630 PROCTOR (R. A.). Rough Ways Made Smooth; Familiar Science Studies; Popular Account of Transits of Venus, 1639-2012; Pleasant Ways in Science; Our Place Among Infinities; Universe of Suns and other Science Gleanings; Mysteries of Time and Space. *Some vols. illustrated.* Together 7 vols. Small 8vo, cloth (6 vols. with top edges gilt).   London and N. Y., 1880-84, etc.

631 PROCTOR. Familiar Science Studies. Small 8vo, cloth.
N. Y., 1882

632 PROSE MASTERPIECES FROM MODERN ESSAYISTS. *Portraits.* 3 vols. small 4to, new cloth, top edges gilt, others uncut.   N. Y., n. d.
Includes Essays by—Irving, Leigh Hunt, Lamb, De Quincey, Landor, Sydney Smith, Thackeray, Emerson, Matthew Arnold, Morley, Helps, Kingsley, Ruskin, Lowell, Carlyle, Macaulay, Froude, Freeman, Gladstone, Cardinal Newman and Leslie Stephen.

633 PUBLIC LIBRARIES in the U. S. A. Part 1. *Illustrations.* Thick large 8vo, sewed, uncut.   Washington, 1876

634 PUNCH. Half a Century of English History, *presented in a series of cartoons from "Punch."* Small 8vo, new cloth.
N. Y., 1884

635 PUNCH. Another copy of the same.

636 PURVES (D. L.). English Circumnavigators—DRAKE, etc. *Maps.* Thick small 8vo, cloth, top edge gilt. London, 1874

637 PUTNAM (George P.). The World's Progress, an Index to Universal History, and a Cyclopædia of Facts, Dates, and General Information. Revised and extended to the present time by FREDERICK B. PERKINS and LYNDS E. JONES. *With portrait.* Thick large 8vo, cloth. N. Y., 1883
The work contains a Chronological and Alphabetical Record of all essential facts in the progress of society from the beginning of history to the present time. It is accompanied by a chart of history representing the rise, revolutions, and fall of the principal empires of the world.

637* PUTNAM. Another copy of the same.

638 PUTNAM'S Art Hand Books. *Illustrated.* 2 vols. square
   small 8vo, new cloth. N. Y., 1885
   > Includes—Sketching from Nature and in Water-Colors, Land-
   > scape and Flower Painting, Drawing in Black and White, Figure
   > Drawing, Water-Color Painting and the Human Figure.

639 RABELAIS (F.). Works. *With numerous illustrations
   by* GUSTAVE DORÉ. Small 8vo, cloth. London, n. d.

640 RACINE. The Suitors, a Comedy, translated by IRVING
   BROWNE. *Front.* 4to, cloth.
   LARGE PAPER. N. Y., 1871

641 RAMBLER and IDLER, by JOHNSON; Adventurer, by
   HAWKESWORTH; and Connoisseur by Mr. TOWN. *Portrait.* Thick large 8vo, cloth. London, 1877

642 RAMSAY (E. B.). Reminiscences of Scottish Life and
   Character. *Illustrated.* Small 8vo, cloth. N. Y., 1884

643 RANKE (L. von). Civil Wars and Monarchy in France.
   12mo, cloth. N. Y., 1854

644 RENAN (Ernest). Recollections of My Youth. Square
   12mo, cloth. N. Y., 1883

645 REPRESENTATIVE AMERICAN AND BRITISH ORATIONS. Together 6 vols. Small 4to, new cloth, top edges gilt, others
   uncut. N. Y., 1884

646 REPRESENTATIVE ESSAYS by IRVING, LAMB, DE QUINCEY,
   etc. *Portrait.* Small 8vo, new half russia, cloth sides.
   N. Y., 1885

647 REYNAUD (L.). Traite d'Architecture, *Paris,* 1875; also
   Holy Bible, *profusely illustrated,* but imperfect. Together 2 vols. Large 4to, sewed.

648 ROBINSON (Solon). Thatsachen für Landwirthe sowie
   für Familienkreis. *Plates.* 2 vols. large 8vo, roan gilt,
   edges gilt (rubbed). Cleveland, 1868

649 RODENBOUGH (T. F.). Afghanistan and the Anglo-Russian
   Dispute. *Maps and other illustrations.* Small 8vo, new
   cloth. N. Y., 1885

650 ROGET (P. M.). Thesaurus of English Words and Phrases.
   12mo, cloth. N. Y., 1883

651 ROLLIN (Charles). Ancient History. 4 vols. small 8vo,
   cloth. N. Y., n. d.

652 ROOSEVELT (Theodore). Naval War of 1812. *Cuts.* 8vo,
   new cloth. N. Y., 1882

653 ROUSSEAU (J. J.). Confessions of. *Illustrated.* Small 8vo, fresh cloth.　　　　　　　　　　　　London, 1875

654 ROUSSELET (Louis). The Serpent Charmer and numerous other Stories, Tales, Adventures, etc. *With 500 illustrations.* 4to, cloth (cover loose).　　　　　London, n. d.

655 ROYCE (Samuel). Deterioration and Elevation of Man through Race Education. 2 vols. 12mo, cloth.
　　　　　　　　　　　　　　　Boston, 1880

656 SABINE (Lorenzo). Biographical Sketches of Loyalists of the American Revolution. 2 vols. 8vo, cloth.
　　　　　　　　　　　　　　　Boston, 1864

657 ST. AUGUSTINE, Florida. Small 8vo, cloth.　N. Y., 1869

658 SAINTE BIBLE, revue par DAVID MARTIN. Thick large 8vo, morocco (rubbed and spotted).　　　　Paris, 1840

659 ST. JOHN (J. A.). Egypt and Nubia. *Illustrated.* 8vo, cloth, uncut.　　　　　　　　　　London, 1845

660 ST. NICHOLAS MAGAZINE. *Profusely illustrated.* 52 parts. Large 8vo, sewed.　　　　　　　N. Y., 1877–84

661 SAMUELS (Edward A.). Ornithology and Oology of New England. *Numerous colored plates and wood-engravings.* Thick 4to, cloth.
　　　LARGE PAPER.　　　　　　　　Boston, 1868

662 SAMUELS. Our Northern and Eastern Birds. *Numerous illustrations, some colored.* Thick large 8vo, cloth, gilt.
　　　　　　　　　　　　　　　N. Y., 1883

663 SAMUELS. Birds of New England. *Illustrated.* Thick 8vo, cloth (damaged in places).　　　Boston, 1883

664 SCHIMMELPENNINCK (M. A.). Theory of the Classification of Beauty and Deformity. *Plates on colored paper.* 4to, cloth (name on title).　　　　　London, 1815

665 SCHLEGEL (F.). History of Literature ; Traces of Travel ; BUTLER's Analogy. Together 3 vols. Cloth.

666 SCHUBERT (G. H. von). Naturgeschichte der Säugethiere. *With 159 different objects in 30 plates.* Small folio, boards (back damaged).　　　　　Esslingen, 1878

667 SCHUCKERS (J. W.). Life and Public Services of CHIEF JUSTICE S. P. CHASE. *Portrait.* Thick 8vo, cloth.
　　　　　　　　　　　　　　　N. Y., 1874

668 SCIENCE LADDERS. *Illustrated.* Thick small 4to, cloth.
　　　　　　　　　　　　　　　N. Y., 1884

669 SCOTT (E. G.). Development of Constitutional Liberty in the English Colonies of America. 8vo, cloth.
N. Y., 1882

670 SCOTT (Sir Walter). Poetical Works, with Life, by WILLIAM CHAMBERS, LL.D. *Illustrated.* 4to, cloth, gilt edges (binding inked somewhat). N. Y., 1883

671 SCOTT. Poetical Works. *Red line borders and illustrated.* Small 8vo, roan gilt, edges gilt. N. Y., 1881

672 SCOTT. IVANHOE, a Romance. EDITION-DE-LUXE. Thick large 8vo, cloth, top edge gilt (cover loose). London, n. d.

673 SCOTT. Ivanhoe and Talisman. *Illustrated.* 2 vols. in 1. Small 8vo, half calf, gilt. N. Y., 1880

674 SCOTT. Waverley Novels—Handy Volume Edition, Vols. 1 to 23 inclusive, also Vol. 25. Together 24 vols. Minimo, cloth, purple edges. London, 1877

675 SCRIBNER'S MONTHLY and Century. *Profusely illustrated.* 29 parts (some duplicates). Large 8vo, sewed.
N. Y., 1878–83

676 SEWARD (William H.). Autobiography of, from 1801 to 1834. *Portraits.* Thick 8vo, cloth. N. Y., 1877

677 SHAKSPERE (William). Works of, revised from the Best Authorities, with Memoir and Essay on his Genius, by BARRY CORNWALL, also Annotations and Introductory Remarks on the Plays by many Distinguished Writers. *Illustrated with numerous engravings on wood from designs by* KENNY MEADOWS. 3 vols. large 8vo, calf gilt, marbled edges. London, *Tyas,* 1843

678 SHAKESPEARE. Poetical Works. *Illustrated and red line borders.* Small 8vo, half russia gilt. N. Y., 1881

679 SHAW (Albert). Icaria, a Chapter in History of Communism. Small 4to, cloth. N. Y., 1884

680 SMALLEY (Eugene V.). History of Northern Pacific Railroad. *Illustrated and maps.* Thick 8vo, cloth.
N. Y., 1883

681 SMITH (Gerritt). Biography of, by O. B. FROTHINGHAM. *Portrait.* Small 8vo, fresh cloth. N. Y., 1879

682 SMITH (Matthew Hale). Successful Folks. *Portraits.* 8vo, cloth. Hartford, 1879

683 SMITH (Philip). Ancient History, from Earliest Records to Fall of Western Empire. *With maps and plans.* 3 vols. 8vo, cloth, uncut. London, 1868

684 SMITH (Roderick H.). Science of Business. *Diagrams.*
Small 8vo, fresh cloth.　　N. Y., 1885

685 SMITH (Sydney). Memoirs, Letters and Essays. 2 vols.
small 8vo, cloth.　　London, n. d.

686 SMITH (Walter). Examples of Household Taste. *Illustrated.* Cloth, top edge gilt (rubbed).　　N. Y., n. d.

687 SMITH (Wm.). Ancient History of the East. *Cuts.* 12mo,
cloth.　　N. Y., 1872

688 SMITH. Smaller Classical Dictionary. 200 *cuts.* Small
8vo, cloth.　　London, 1876

689 SMOLLETT (Tobias). Works, with Life. *Portrait and
vignette title.* Large 8vo, cloth, top edge gilt.
　　Boston [Edinburgh], *n. d.*

690 SMOLLETT. Adventures of Roderick Random. *Vignette
cuts on titles.* 2 vols. minimo, half morocco, top edges
gilt.　　Chiswick, *C. Whittingham*, 1823

691 SMYTH (William). Lectures on Modern History. 2 vols.
8vo, cloth, uncut.
　　London, *W. Pickering, by C. Whittingham,
Chiswick,* 1841

692 SONGS OF THE SPIRIT, edited by ODENHEIMER and BIRD.
Thick 4to, cloth, gilt edges.　　N. Y., 1871

693 SPECTATOR (The), with a Historical and Biographical
Preface, by CHALMERS. *Portrait.* 8 vols. small 8vo,
parchment, top edges gilt, others uncut.　　N. Y., 1883

694 SPENCER (Herbert). First Principles and Data of Ethics.
2 vols. small 8vo, cloth.　　N. Y., 1875-79

695 STACK (Edward). Six Months in Persia. *With 7 maps.*
2 vols. 8vo, cloth.　　N. Y. [London], 1882
　　"He gives his readers an entertaining narrative, * * * and
what seems to be a just idea of the Empire and its conditions.
* * * He has useful chapters on the geography of the country and
its land-revenue system. * * * His book certainly is instruc-
tive and entertaining in a high degree."—*Congregationalist.*

696 STACK. Another copy of the same.

697 STANLEY (T. Lloyd). Future Religion of the World. 8vo,
new cloth.　　N. Y., 1884

698 STEELE (J. W.). Cuban Sketches. Small 8vo, fresh cloth.
　　N. Y., 1881

699 STEPHEN (Leslie). History of English Thought in the 18th
Century. 2 vols. 8vo, cloth.　　N. Y. 1881

700 STEPHEN. Science of Ethics. 8vo, fresh cloth.
N. Y. [London], 1882

701 STEPHEN. Another copy of the same.

702 STEPHENS (A. H.). Constitutional View of War between the States. *Portraits and view.* Vol. 1. Large 8vo, cloth. Phila., 1868

703 STEPHENS in Public and Private, by CLEVELAND. *Portraits and views.* 8vo, cloth. Phila., 1866

704 STEPHENS. Life of, by JOHNSTON and BROWNE. *Portrait and view.* 8vo, cloth. Phila., 1878

705 STEPHENS (John L.). Travels in Central America, Chiapas and Yucatan. *Map and illustrations.* 2 vols. large 8vo, cloth (somewhat foxed). SCARCE. N. Y., 1855

706 STEPHENS. Travels in Egypt, Arabia Petræa, etc. *Illustrations.* 2 vols. 12mo, cloth. N. Y., 1854

707 STERNE (Laurence). Works, with Life. DARLEY's *illustrations.* 8vo, cloth. Phila., 1873

708 STEVENS (John L., *LL.D.*). History of GUSTAVUS ADOLPHUS. *Portrait.* Thick 8vo, new cloth. N. Y., 1885

709 STIELER (Adolf). Hand Atlas, uber alle theile der erde und uber das weltgebaude. *Colored maps.* Thick folio, half russia, cloth sides. Gotha, 1866

711 STILLE's History U. S. Sanitary Commission, *Phila.*, 1866; HOPKINS' View of Slavery; etc. Together 5 vols. Cloth.

712 STONE (William L.). Reminiscences of Saratoga and Ballston. *Illustrated.* Small 8vo, cloth. N. Y., 1880

713 STORIES for Telling to Children, by "PRUDENTIA." *Illustrated.* 4to, illuminated cloth. (2 copies.) N. Y., n. d.

714 STOTHERT (James). French and Spanish Painters—a Critical and Biographical Account of the Most Noted Artist of the French and Spanish Schools. *With 20 illustrations on steel after* MURILLO, GOYA, FORTUNY, DE LA ROCHE, PRUD'HON, MEISSONIER, INGRES *and other Famous Masters, etched by* FLAMENG, RAJON, CAREY, HEDOUIN, GAILLARD, *etc.* Thick 4to, cloth, gilt edges.
London, n. d.

715 STRAUSS (D. F.). Life of Jesus Critically Examined. 3 vols. 8vo, cloth, uncut. London, 1846

716 SUE (Eugene). Mystères de Paris. 6 vols. 8vo, half calf.
Paris, 1842

717 SUE. Le Juif Errant. 9 vols. small 8vo, half roan, uncut. Paris, 1844–45
718 SWEETSER (M. F.). Handbook of Boston Harbor. *With
200 illustrations.* Small 4to, cloth, gilt. Cambridge, 1883
719 SWIFT (Jonathan, D.D.). Works and Memoir by Roscoe,
with Copious Notes and Additions. 6 vols. 12mo, cloth.
N. Y., *n. d.*
720 SWIFT. Works, with Life and Notes. 2 *portraits.* Large
8vo, cloth, top edge gilt. Brooklyn and N. Y., n. d.
721 SWINBURNE (Algernon Charles). Studies in Song and
MARY STUART. 2 vols. in 1. Thick 12mo, half calf, gilt.
N. Y., 1880–81
722 SWINBURNE. MARY STUART, a Tragedy. 12mo, cloth, top
edge gilt. N. Y., 1881
723 SWINBURNE. Tristram of Lyonesse and other Poems.
Small 8vo, sewed. London, 1882
724 SWINBURNE. Midsummer Holiday and other Poems.
Small 8vo, cloth. N. Y., 1884

725 TACITUS and HOMER's Iliad, literally translated. *Portrait.* 3 vols. 12mo, cloth. N. Y., 1856–75
726 TAINE (H. A.). Ancient Régime. 12mo, cloth.
N. Y., 1876
727 TALBOT (C. R.). Parlor Comedies. *Illustrated.* 4to,
cloth. Boston, 1883
728 TAMENAGA SHUNSUI. The Loyal Ronins, an Historical
Romance translated from the Japanese by E. GREEY and
SHIUICHIRO SAITO. *Illustrated by* KEI-SAI YEI-SEN *of
Yedo.* 4to, new cloth, gilt. N. Y., 1884
729 TAUSSIG (F. W.). Protection to Young Industries in
America. 8vo, fresh cloth. N. Y., 1884
730 TAYLOR (Bayard). India, China and Japan. Small 8vo,
cloth. N. Y., 1884
731 TAYLOR (Richard, *Confederate General*). Destruction and
Reconstruction. 8vo, new cloth, uncut. Edinburgh, 1879
732 TENNYSON (Alfred). Poetical Works. *Numerous illustrations.* Large 8vo, cloth, gilt edges. Boston, n. d.
733 TENNYSON. Complete Works. *Red line borders and illustrated.* Small 8vo, half russia, gilt (cover loose).
N. Y., 1881

734 TENNYSON. The Lady of Shalott. *Chromolithographic illustrations by* HOWARD PYLE. 4to, cloth, gilt.
N. Y., 1881

735 TENNYSON. Another copy of the same.

736 TENNYSON. Dream of Fair Woman. *Illustrated.* 4to, cloth, gilt edges.
Boston, 1884

737 THACKERAY (William Makepeace). Works, People's Edition. *With* 325 *illustrations by the author,* DU MAURIER, CRUIKSHANK, LEECH, MILLAIS, BARNARD *and others.* 20 vols. in 10. Thick small 8vo, new cloth.
Boston, *Estes & Lauriat,* 1882–83

738 THACKERAY. The Newcomes. *With* DOYLE's *illustrations.* Small 8vo, half calf, gilt.
N. Y., n. d.

739 THACKERAY. Barry Lyndon, Great Hoggarty Diamond, Character Sketches, Men's Wives, etc. *With author's illustrations.* Small 8vo, half russia, marbled edges.
N. Y., 1880

740 THIERS (A.). Histoire de la Revolution Francaise. *Steel engravings, etc.* 10 vols. 8vo, half morocco, gilt (some vols. slightly foxed).
Paris, 1839

741 THIERS. Histoire du Consulat et Empire. Vols. 1 to 4 inclusive. 8vo, half calf, cloth sides (a few pp. stained).
Paris, 1845

742 THIERS. History of French Revolution. *Plates.* 5 vols. 8vo, half calf (foxed somewhat).
London, 1838

743 THIERS. History of the French Revolution. *Illustrated.* 4 vols. in 2. 8vo, cloth.
N. Y., 1874

744 THOMSON (Richard). HISTORICAL ESSAY ON THE MAGNA CHARTA OF KING JOHN, to which are added various Old Charters, with Translations and Notes. *Every page surrounded by an elegant and differently designed heraldic border.* Thick 8vo, cloth, uncut (a few pp. slightly foxed).
London, *Major,* 1829

A splendid copy with *India paper proof* title, and illustrated with upwards of 600 heraldic and chivalric devices, 69 cuts of monumental effigies; also seals, emblems, etc. All within exquisite ornamental borders.
"A book as beautifully and appropriately adorned as it is elaborately and learnedly compiled."—SOUTHEY.

745 THOUSAND AND ONE GEMS of English Poetry. *Red borders and illustrated.* Small 8vo, half russia, gilt. N. Y., 1882

746 Thwing (Charles F.). American Colleges. Small 8vo, cloth.   N. Y., 1883

747 Tilton (Theodore). Tempest Tossed, *portrait;* Swabian Stories by Tilton. Together 2 vols. 12mo, cloth.
.   N. Y., 1882–83

748 Timbs (John). Curiosities of History ; Second Series of Science Curiosities; and Popular Errors Explained· *Fronts.* 3 vols. small 8vo, cloth, uncut.   London, 1856–60

749 Tocqueville (Alexis de). Republic of U. S. A. Thick square 8vo, cloth.   N. Y., n. d.

750 Tocqueville. Old Régime and the Revolution. 12mo, cloth.   N. Y., 1856

751 Token of Friendship; Rose of Sharon; Flora's Interpreter; etc. Together 12 vols. Full leather.

752 Topics of the Time—Studies in Literature, Social Problems and Studies in Biography. 3 vols. small 8vo, new cloth.   N. Y., 1883

753 Tourist's Guide-Book to United States and Canada. *Maps and illustrations.* Small 8vo, roan, gilt edges. N. Y.., 1884

754 TRANS-ATLANTIC NOVELS, *i. e.:*—Gautier's Captain Fracasse; Dingelstedt's The Amazon; Peard's Mother Molly; Boisgobey's Lost Casket, and The Golden Tress; Crawfurd's World We Live In; Macquoid's Her Sailor Love and Esau Rinswick; Sime's King Capital; My Trivial Life, 2 vols.; Halevy's Abbé Constantine; Edwardes' Eleventh Hour ; Muir's Lady Beauty ; Murray's Joseph's Coat; John Barlow's Ward; Fenn's Vicar's People; Rochefort's Mlle. Bismarck. Together 18 vols. Square 12mo, fresh uniform blue cloth.
                                                         N. Y., 1880–84

755 Treaty of Washington Papers. 6 vols. 8vo, cloth.
                                                Washington, 1872–74

756 TURNER (Mrs. C. H.). The Floral Kingdom, its History, Sentiment and Poetry. *Illustrated and carmine borders.* 4to, cloth (rubbed).   Chicago, 1877

757 Turner. Cyclopedia of Practical Floriculture. *Fancy initial letters and rubricated borders.* 4to, cloth, gilt edges (side damaged).   N. Y., 1884

758 Tupper (Martin F.). Poetical Works. *Portrait and red line borders.* Small 8vo, half calf, gilt.   N. Y., 1882

759 Tupper. The Same Edition. Half russia, gilt.

760 **U.** S. HYDROGRAPHIC REPORTS. *Maps, plates.*
etc. 20 vols. 8vo, cloth. Washington, *v. d.*
Includes—Gorringe's Rio de la Plata, 1875 ; Labrosse's Atlantic
Ocean, 1873, and Pacific Ocean, China Seas, etc., 1875 ; Kerhal-
let's Pacific Ocean, 1869, Indian Ocean, 1869, and Atlantic Ocean,
1870 ; South Coast of England, 1869 ; Gorringe and Cheney's
West Coast of Africa, 2 vols. 1873-75 ; LeGras's Mediterranean,
1870 ; Wyman's Gulf of Cadiz, 1870 ; Kropp's Red Sea, 1872 ;
Gorringe's Coast of Brazil, Vol. 1, 1873 ; Totten and Schroeder's
Bay of Biscay, 1876 ; Gorringe's Mediterranean, 1875 ; Kerhallet's
Azores, 1874 ; Kerhallet and Le Gras's Madeira, etc., also Cape
Verde, 1873-74 ; Totten's Coasts of Spain and Portugal, 1874 ;
Davey's California and Mexico Coasts, 1874.

761 U. S. GEOGRAPHICAL AND POLITICAL ATLAS, or U. S. Land
Survey Maps, 23 pieces; Battle of Gettysburg Maps, 3
pieces. (26 pieces)

762 **V**ALLEE (Leon). Bibliographic des Bibliographies.
Thick large 8vo, sewed, uncut. Paris, 1883
The most complete catalogue of catalogues yet made. It is in-
valuable to Librarians, and treats as extensively of English and
American Bibliography as that of Continental Europe.

762* VALLEE. The same. Another copy. ·

763 VIENNA INTERNATIONAL EXHIBITION, 1873. Reports of
U. S. Commissioners. *Illustrated.* 4 vols. 8vo, sheep.
Washington, 1876

764 **W**ARNER (S. *and* A.). Little American Stories and
Sketches. Large 8vo, cloth. West Point, 1863

765 WARREN'S Now and Then; HEAD's Faggot of French
Sticks; DANIEL BAKER's Life and Labors; DUMONT'S
Lady's Oracle; NOTT's Sketches of the War; DALTON's
John Chinaman; etc. Together 10 vols. Cloth.

766 WEBSTER (Daniel) *and* CLAY (Henry). Obituary Addresses. *Portraits.* 2 vols. 8vo, cloth.
Washington, 1852–53

767 WEISE (A. J., *M.A.*). The Discoveries of America to the
Year 1525. *Maps and fac-similes.* Thick 8vo, cloth,
top edge gilt, others uncut. N. Y., 1884

768 WELLS (D. A.). Our Merchant Marine. *Front.* Small
8vo, fresh cloth. N. Y., 1885

769 WESTMINSTER REVIEW, from January, 1863, to October, 1874, inclusive. Together 36 parts. Large 8vo, sewed.
N. Y., 1863–74

770 WHITMAN (Walt). Leaves of Grass. *Portrait.* 12mo cloth. Boston, 1860–61

771 WHITMAN. Specimen Days and Collect. 12mo, cloth.
Phila., 1882–83

772 WHITMAN. Another copy of the same.

773 WHITTIER (J. G.). Mabel Martin. *Illustrations.* 8vo, cloth, gilt edges. Boston, *Riverside Press*, 1881

774 WHITTIER. Life, Genius and Writings of, by KENNEDY. *Portrait.* Small 8vo, cloth. Boston, 1882

775 WIELAND. Oberon, translated by SOTHEBY. 12 *plates by* HEATH *and others, after* FUSELI. 2 vols. small 8vo, calf (titles cut and binding damaged). London, *Bulmer*, 1805

776 WILLIAMS (G. W.). History of the Negro Race in America from 1619 to 1880; Negroes as Slaves, as Soldiers and as Citizens. *Portrait.* 2 vols. large 8vo, cloth.
N. Y., 1883

777 WILLIAMS (W.). OXONIA DEPICTA, sive Collegiorum met
• Dularum in Inclyta Academia Oxoniensi. *Illustrated with* 65 *plates of views, plans, maps, etc.* Large folio, old sheep (binding broken, etc.—three pages mounted).

778 WILSON (Andrew). Chapters on Evolution. *With* 259 *illustrations.* Small 8vo, cloth, totally uncut. N.Y., 1883

779 WILSON (John). Recreations of CHRISTOPHER NORTH. *Portrait.* Large 8vo, cloth (one p. inked). N. Y., 1860

780 WOLTMANN (Alfred) *and* WOERMANN (Karl). History of Ancient, Early Christian and Mediæval Painting, from the German, and edited by SIDNEY COLVIN. Cloth, top edge gilt, others uncut. N. Y., 1880

781 WOMAN QUESTION in Europe. Original Essays edited by STANTON, with Introduction by COBBE. Large 8vo, cloth.
N. Y., 1884

782 WOOD (D. G.). Popular Natural History. *Profusely illustrated.* Small 4to, cloth. N. Y., 1884

783 WOOLFOLK (L. B.). The World's Crisis. 8vo, cloth.
Cincinnati, 1868

784 WRIGHT (R. J.). Principia, or Basis of Social Science. Thick 8vo, cloth. Phila., 1876

785 ZAMBA (*African Negro King*). Life of, and Slavery in South Carolina. *Front.* Small 8vo, half roan. London, 1847

www.ingramcontent.com/pod-product-compliance
Lightning Source LLC
Chambersburg PA
CBHW022016190326
41519CB00010B/1542